Humane Infrastructures

AF698314

metaLABprojects

The metaLABprojects series provides a platform for emerging currents of experimental scholarship, documenting key moments in the history of networked culture, and promoting critical thinking about the future of institutions of learning. The volumes' eclectic, improvisatory, idea-driven style advances the proposition that design is not merely ornamental, but a means of inquiry in its own right. Accessibly priced and provocatively designed, the series invites readers to take part in reimagining print-based scholarship for the digital age. www.metalab.harvard.edu

Series Editor

Jeffrey T. Schnapp

Humane Infrastructures

Patrik Svensson

The MIT Press
Cambridge, Massachusetts
London, England

The MIT Press
Massachusetts Institute of Technology
77 Massachusetts Avenue, Cambridge, MA 02139
mitpress.mit.edu

© 2025 Massachusetts Institute of Technology

This work is subject to a Creative Commons CC-BY-NC-ND license.

This license applies only to the work in full and not to any components included with permission. Subject to such license, all rights are reserved. No part of this book may be used to train artificial intelligence systems without permission in writing from the MIT Press.

(cc) BY-NC-ND

The MIT Press would like to thank the anonymous peer reviewers who provided comments on drafts of this book. The generous work of academic experts is essential for establishing the authority and quality of our publications. We acknowledge with gratitude the contributions of these otherwise uncredited readers.

This book was set in Minion Pro by Westchester Publishing Services. Printed and bound in the United States of America.

Library of Congress Cataloging-in-Publication Data

Names: Svensson, Patrik, author.
Title: Humane Infrastructures / Patrik Svensson.
Description: Cambridge, Massachusetts : The MIT Press, [2025] |
Series: metaLABprojects | Includes bibliographical references and index.
Identifiers: LCCN 2024028676 (print) | LCCN 2024028677 (ebook) |
ISBN 9780262542128 (paperback) | ISBN 9780262362290 (epub)
Subjects: LCSH: Digital humanities—Social aspects. | Humanities—Social aspects. |
Infrastructure (Economics)—Social aspects.
Classification: LCC AZ105 .S884 2025 (print) | LCC AZ105 (ebook) |
DDC 363—dc23/eng/20241014
LC record available at https://lccn.loc.gov/2024028676
LC ebook record available at https://lccn.loc.gov/2024028677

10 9 8 7 6 5 4 3 2 1

EU product safety and compliance information contact is: mitp-eu-gpsr@mit.edu

Contents

Preface

On April 27, 2023, the Swedish government announced a decision to appoint 238 external board members for thirty Swedish universities.[1] Candidates for each university had been nominated by two respected civic members elected by the government, including county governors and former university presidents. There is a procedural regularity to these decisions. This time, things were different, however, as the regular tenure of three years had been more than halved at short notice extremely late in the process. The rationale given for the decision was the geopolitical situation and the need for security policy competence on the boards. A heated national debate ensued, including a protest letter to government from essentially all university presidents in the country, where this rationale was questioned.[2] The presidents claimed that the boards are strategic, not operative, and that if the government wants to make sure universities fulfill security policy requirements, there are other, better ways. Also, there would be at least seventeen months until the boards could be changed, which seemed slow if there was an urgent need. Two weeks later, the Swedish Association of University Teachers and Researchers went as far as to claim that Swedish universities had been kidnapped by the right-wing governing bloc.[3]

At the time of writing, it is unclear what this decision will lead to, if anything, but control of Sweden's largest state-funded sector is on the table. What if we were to take the geopolitical security-driven argument one step further in a thought experiment? While not very likely, it is not unforeseeable that the next step in this development would be to classify higher education as critical infrastructure—assets that provide "functions necessary for our way of life," such as energy, water, and communications, in need of special protection from threats.[4] There is a close relationship between protecting and controlling, and emergencies—such as perhaps a perceived threat to the integrity of higher education—can sometimes require extreme measures. Through framing higher education as a critical infrastructure under threat, such measures could potentially be rationalized.

Stephen Collier and Andrew Lakoff trace critical infrastructure protection back to a longer history of preparedness and risk management, including biopolitical "population security" (a concern with public health, social insurance, etc.) in the nineteenth century where collective life was gradually reimagined in terms of a distribution of possibilities and risks—partly driven by collection of and access to large quantities of data on factors such as mortality and level of health ("demographics")—and a growing concern in the twentieth century with the vulnerability of interlinked systems such as transportation, electricity, and water ("vital systems security").[5] These large systems, vital to collective life, were increasingly seen as susceptible to unpredictable and far-reaching breakdowns and risks, including wars, industrial accidents, and labor strikes.[6]

Following this development, the concept of "critical infrastructure" emerged in the mid-1990s in the US, partly triggered by the Oklahoma City bombing (which targeted a federal building) and by an increasing awareness of the risk of cyber threats and information warfare, and was institutionalized through the Presidential Commission on Critical Infrastructure Protection set up by President Clinton in 1996.[7] Such concerns, of course, were accelerated through 9/11, and critical infrastructure protection and systems vulnerability continue to be very current topics in the wake of the COVID-19 pandemic, climate crisis, and the war in Ukraine.

There are now many national and pan-national critical infrastructure protection frameworks, which all have their own genealogies.[8] Critical infrastructure is negotiated through such frameworks and is by no means a set category. An analysis of US critical infrastructure protection describes definitions as "fluid" and extending over time, even to monuments and icons.[9] In a report on European Union and Swedish critical infrastructure protection, we are told that deciding what counts as critical infrastructure can be challenging.[10] Such decisions carry considerable weight (What would it mean if higher education was made into a critical infrastructure?) and are not necessarily unproblematic, especially as they are linked to national and pan-national security.

For example, critical infrastructure protection can be used to reduce or stop dissent or to reduce participation in societally important decisions, as when pipelines are installed across Indigenous land or without engaging communities.[11] Indigenous actors in Canada have been classified according to the threat level they pose to critical infrastructure as part of national security policing.[12] Could we imagine a threat-level system for classifying academics and research areas if universities were critical infrastructures? The frameworks of critical infrastructure protection make possible such and other mechanisms through what Julian Reid describes as "increasingly invasive techniques of governance."[13] While there is no question that infrastructures can be under threat and that some need to be protected, critical infrastructure is part of complex landscape of fear or even dread, securitization, protectionism, globalization, technocentrism, and vested stakeholder interests (including governments and the security industry).[14] Collier and Lakoff assert that the underlying security threats are neither just out there ("self-evident facts") nor freely constructed ("simply imagined").[15] In Sweden, security concerns suddenly became an issue for appointing university boards, and while it would certainly be a stretch to categorize Swedish universities as critical infrastructure, the question is how much of a stretch.

Critical infrastructure serves as a reminder of the central role we have given to infrastructure. It is what we need to securitize to protect our way or life and the "health, safety, security or economic well-being of citizens or the effective functioning of governments."[16] Claudia Aradau argues that the securitization of critical infrastructures assumes that infrastructure is "the very foundation of society."[17] This is an equally remarkable and self-evident assumption: self-evident (although intriguing) if we consider the fabric of society to be infrastructural and infrastructure as a given that undergirds our society, and remarkable if we see infrastructure as negotiated and as a reflection of specific values, assumptions, and choices, as discussed above. In either case, if infrastructure is what makes the world, we need to pay careful attention, especially if we are interested in changing that world.

The world before the 1980s—when "infrastructure" as a concept and idea gained more currency in many countries—was less infrastructural. It is not that we have not relied on technology and supportive systems for a very long time, whether aqueducts, libraries, roads, or railways, but rather that "infrastructure" increasingly became a way of thinking about the world. This way of thinking was not entirely new, sudden, or all inclusive, and for example, it relied heavily on a modernist conception of rationality, on technology, and on "public works" as a frame.[18] It also co-emerged with a series of developments following World War II, including the rebuilding of European defense infrastructure, globalization, the deregulation of public works, the fight against terrorism, marketization, the expansion of higher education, new public management, and digitalization.

"Research infrastructure," one example of this development, came into being from the mid to late 1990s in Canada and Europe (and elsewhere) as a means to provide shared facilities and tools to researchers. It was driven by ideas about research excellency and innovation, advances in technology, global competition, and the European project. The focus was science and technology instruments of considerable scale and cost, such as telescopes and synchrotron radiation facilities. Over time, research infrastructure became a way to organize higher education—especially in Europe—through nationally sanctioned conditions, road maps, distribution of resources, and committees at every level. This ongoing development has been entwined with the expansion of the sector, the growth of new public management, and parameters such as control, standardization, accountability, and measurability.

Accordingly, complex frameworks with multiple lineages have arisen around concepts such as "research infrastructure" and "critical infrastructure," where resources and priorities are negotiated as well as what gets included and excluded. In such frameworks, infrastructure is neither merely an encoding of directionality or policy nor a singular force or logic in its own right. Furthermore, infrastructure is not just about social practice and governance but also about material objects, things.[19] When I think of infrastructure, I think of things such as road systems, bridges, visualization equipment, and search engines—things embedded in a sociotechnical context.

Essentially all global challenges and complex problems, including climate change, are infrastructural in one way or another. Fossil-fuel extraction is infrastructural. Legacy infrastructure such as road and aviation systems have a large carbon footprint. The projection models for climate change are computationally and infrastructurally embedded. Measurements collected over centuries using different types of technologies and mechanisms are digitally aligned. Countermeasures and "solutions" are often infrastructural in nature, such as "climate-resilient infrastructure," solar power, and geoengineering. Infrastructure is part of both the problem and the solution. How possible solutions are imagined and built, then, partly becomes an infrastructural question, frequently falling back to a conventional, if well-meaning, view of infrastructure.

On one level, climate change is a scientific and technological challenge. At the same time, viable responses must take into account how complex issues are embedded culturally and socially, which is relevant to understanding the problem in question as well as devising strategies for how to respond to it.[20] Some of these responses will likely require us to shift the conception or frame of what infrastructure is and can be, including making sure to care for our planet, embracing communities, supporting small as well as large solutions, considering cultural and societal structures, and invoking critical and historical knowledge as well as our collective imagination.

Changing infrastructure requires working across different levels and modes of engagement, including underlying ideas and frames, frameworks, and material manifestations. A green city cannot be created through blocking one road off or producing city municipality documents specifying a commitment to greening the city. Also, because infrastructure is templated, change becomes more difficult the more one moves away from the technological, large size, and instrumental core of infrastructure as an idea. In this sense, a "green city" seems less infrastructural than a "smart city," and oil pipes more like infrastructure than Indigenous land.[21] Nevertheless, the materiality of infrastructure cannot just be a second thought or a simple manifestation of ideas. Often, the material layer can be or manifest the bringing together of perspectives. The making of infrastructure is a material-conceptual-discursive task.

Infrastructure is regularly portrayed as having agency, a tendency driven by its technological nature, and although we should be skeptical about such framings, it is true that "infrastructure" enables and hinders directionality, in a sense setting up our freedom of movement. It also makes us anticipate and align our work and thinking with given templates, whether a big corporate, tech-driven model of urban development or a science model for research infrastructure. Such templates manifest infrastructural thinking—for example, infrastructure as big, technological, and instrumental—which has come to seem natural, even when making things inadvertently worse. I argue that we need to open up a space for problematizing and challenging such infrastructural thinking. Doing things differently may require us to push infrastructural models or come up with different models or frames, such as “pop-up cycling infrastructure” (challenging the longevity and stationary nature of infrastructure) or “toxic infrastructure” (challenging the functional and benevolent nature of infrastructure).[22]

What kind of infrastructure do we need then? Building on the work of others, I suggest that we need infrastructure that is informed by humanistic and humane knowledge as well as technological, scientific, financial, and other knowledge. Infrastructure must be humane, civil, and community based as well as part of technological, governmental, and corporate systems. It needs to be imaginative and experimental as well as stable and standardized. Such a shift involves far-reaching engagement across knowledge traditions and practice. This engagement must be reasonably generic and not just restricted to specific issues or competencies. We need infrastructure—or capacity—to mobilize humane and humanistic knowledge at some scale in order to respond collaboratively to societal challenges and complex problems.[23] Infrastructure is by no means everything, but it has become a logic or order of sorts—at least for the time being—and there is space to act. If infrastructure got us to where we are, I suggest that infrastructure can help take us to where we need to go.

Introduction

HUMlab, Umeå University, circa 2010. I used to seat visiting delegations to the lab in front of our large back-projected screen, which was an impressive piece of technology. Occasionally, I would take important visitors to the projection room, which certainly aligned with John Durham Peters's apt notion that infrastructures are systems difficult to carry, including airports and electrical grids.[1] The projector alone was huge and noisy, and it was circumvented by mirrors and various secondary technologies. However, in the early 2010s, although I had started to write about cyberinfrastructure, I would have been unlikely to frame the lab in terms of infrastructure when talking about it and presenting it.[2] Between now and then, the world has become more infrastructural.

People would gravitate toward the seats at the back, distancing themselves from me and the big screen. The talking would stop at some point, especially if I emphasized my presence as the speaker and host through straightening up, crossing my hands in front of me, and looking somewhat intently at them. Once people stop talking, there is no going back. The expectations and roles are crystal clear and were made even more pronounced by the cover slide on the screen behind me. The model of the lecture hall with the teacher up front with a screen behind them goes back hundreds of years, and we have learned how to behave and what to expect.[3]

What I just described is an infrastructural setup manifested not only through the space, screens, software, and seating arrangements but also through the format, expectations, and our bodies. We expect to sit down and be largely silent when a talk or presentation is being delivered. Slides are shown one at a time on a single screen in a familiar pattern. These elements, however, are not fixed, even if they are often taken for granted. Imagine the speaker taking a seat among the participants to initiate a conversation, thus reducing the barrier between these roles, or not using slides at all, unfettering both speaker and audience from the screen and the narrative logic of presentation software.

I largely stopped seating visiting delegations, not only because I found the model passive and static (if comfortable) but also because the idea, simple as it was, emerged in and because of the infrastructure we had built and the community around it. As part of a privileged university setup, we had ten smaller screens at eye height placed peripherally along the outer edge of the same large, multifunctional space (about 2,900 square feet) that featured the large back-projected screen. Johan von Boer, one of our systems and technology experts, implemented a browser plugin that made it possible to fill the screens in the space dynamically through a browser window on a laptop with one tab for each screen. This tool made it possible to create compositions easily and try things out as well as to change content on the fly. In a later development, compositions could be visualized and tested in a web-based 3D rendering of the lab, principally created by technician and 3D expert Mattis Lindmark.

As the infrastructure changed, so did the presentation situation when we had visiting delegations. I would now walk up to one of the peripheral screens. Sometimes, I had to shout at people to come to me or wave, but soon they gathered around me, standing, now much closer than they would have been in the seated configuration, and at the same level as me.[4] It was mostly me talking but with considerably more dialogue and general ease. I had laid out visuals on the screens as part of the story, and we walked from one to the other, seemingly more active than in the earlier setup. Unlike the single-screen paradigm where each slide disappears with the new one, there was retention, as the narrative was still present and referenceable in the space through the visuals as a kind of memory palace.[5]

This interplay between infrastructure and ideas was central to the operation of the lab. There was tension, often productive, in challenging expectations and frames associated with technological systems. For example, I had to insist on keeping the peripheral screens apart when planning the space described above, as there was a will to join them or at least connect them visually through painting lines along the walls. I resisted the ideas of screen-based immersion and scientific visualization walls. The setup—especially the "shiny" technology—could be provocative, as it contradicted both assumptions of humanistic knowledge production and technology-driven logic.

Our strategy was to use infrastructure to get people together around important issues across epistemic traditions, especially concerning challenges to do with knowledge production, and, when appropriate, use and design infrastructure to enact change through an experimental, practice-based, and curiosity-driven approach. Just as any infrastructure is institutionally and organizationally embedded, so were the lab and many of the issues we engaged, and the coming together of the institutional, political, and infrastructural—also beyond knowledge production—is an important part of the story of this book, as is the will to challenge and change not only specific infrastructures but also infrastructural frames, whether civic or academic, as necessary. Crucially, making better or more effective infrastructures is a matter not "only" of attempting to fix problems but also of imagining a more humane society. We need new and realigned partnerships, collaborations, processes, tools, and approaches to respond to the challenges and problems we are facing globally as well as locally, whether it be climate change, urban development, biased software, or unjust systems in society and the academy.

The prototypical referent of "infrastructure" is systems and services that are foundational to society in one way or another, such as roads, electricity, ports, and water systems. In terms of their properties, they tend to be material, systemic, technological, standardized, costly, and "heavy." We also tend to think of infrastructure as long term and reasonably stable. Even if a bridge collapses, the road system as infrastructure and idea is usually not compromised. We will normally expect the bridge to be rebuilt. However, the frameworks and ideas associated with infrastructure can change over time. Roads for cars as an indisputable infrastructure in urban centers may change as a result of a systematic move toward green, slow, pedestrian-friendly cities.[6] Such a shift would likely require a range of changes in policy, regulations, other infrastructures, and "transportation infrastructure thinking" as well as in the material infrastructure.

Today, infrastructure is high on the agenda. Many governments, including in the Netherlands, Japan, Australia, and Sweden, now have a minister of infrastructure.[7] Most of these ministry positions came into being around or after 2000, sometimes as a result of a realignment or extension of earlier cabinets that were pursuing goals or missions that were aligned with what we call "infrastructure" now. In higher education, research infrastructure has become a key priority for many countries. For example, the State Secretary to the Minister for Higher Education and Research in Sweden claimed in 2017 that infrastructure is what takes the most time for the Ministry of Education and Research to handle.[8] The potential gain, however, is taken to be substantial. Cyberinfrastructure in the US has been described as a radical "phase change" with nonlinear effects as part of a long-term shift from an industrial to a knowledge economy.[9] There is often such visionary discourse associated with new technology and infrastructure, as exemplified by "smart cities"' and artificial intelligence (AI).[10] On the other hand, legacy infrastructure, such as roads, bridges, and water systems, is often framed in terms of disrepair and "infrastructure deficit." We have come to understand our world in terms of such narratives as part of a broad societal, political, and material development.

Alongside the conceptual expansion of the term, there has also been an increased critical interest in infrastructure. The uptake of "infrastructure" analytically prompted a 2017 science and technology conference panel to ask why the concept is so productive and what the limitations might be.[11] In a 2018 Modern Language Association panel, it was suggested that "infrastructure" could give the same kind of purchase as "culture" did for Stuart Hall and Raymond Williams, and in a 2021 special issue of the journal *Cultural Studies*, it was argued that "infrastructure constitutes a model of control over meaning and material resources."[12] These statements speak to the critical purchase found in infrastructure, not just as an object of study but also as a disciplinary perspective and critical keyword.

This book starts out by exploring the rise and expansion of the idea of infrastructure—the making of infrastructure—with a particular focus on its conditioning, malleability, and flexibility. How has infrastructure been presented, negotiated, aligned, and changed? Research infrastructure, which came about as a concept in the mid-1990s, serves as a primary case study. This story gets told through a peripheral player in the research infrastructure landscape: the humanities. Research infrastructure was built around science and engineering,[13] and focusing on an outsider allows a different and arguably richer story about the making and negotiation of (research) infrastructure.

This critical awareness of how infrastructure has been shaped and negotiated over time can be turned into a generative condition for simultaneously critiquing, reimagining, and reshaping infrastructure. How can we make infrastructure better and more humane? I argue that we need sustainable and comprehensive ways of mobilizing humanistic, critical knowledge—broadly interpreted—to respond collaboratively to challenges and problems. A central concern becomes where to find such knowledge and how to "tool" it, as well as how to bring together people, communities, and stakeholders in constructive contexts where conditions, frames, and actual infrastructures can be shifted. Ultimately, this is a question of building capacity for making more humane infrastructures.

I take "humane" to refer to approaches that are informed by care and compassion for others, including what can be described as more-than-human actors, animate or inanimate. Humane infrastructures start out from this sensibility and are efficient, life affirming, and not driven primarily or only by technology, standardization, and metrics. There is a significant difference between a humane city and a smart city.[14] I use "humanistic" to refer to knowledge across a range of disciplines that are proximal to the humanities as an academic endeavor. These include a set of emerging areas—such as urban humanities, environmental humanities, disability studies, and food studies—that often emphasize more-than-human perspectives, an active humanism, and infrastructural engagement.

Humane Infrastructures is intended to be useful and relevant to anyone inside and outside the academy interested in how the world—particularly higher education—became infrastructural, how infrastructural frameworks are shaped (exemplified through "research infrastructure"), how such frameworks or regimes can be challenged and changed, and how we can create capacity to collaboratively imagine and build humane infrastructures and a better society. The humanities are important to this vision but as an anchor point for conversation and change within an outward-facing university rather than as an institutional structure, which means that the book is contingent on interaction with a range of fields typically placed outside the academic humanities—including design, infrastructure studies, information studies, and queer studies—and interested parties from beyond the university.

I include some brief, crosscutting perspectives, and positions on infrastructure below—partly reiterating some key points—as a lead-in to the story presented in this book:

- *Infrastructure is not fixed.* When the word "infrastructure" was introduced in the media in the early 1950s in many languages, it was accompanied by considerable uncertainty and hesitancy. In one of the first uses of the term in the Swedish Press, *Dagens Nyheter* reports that the most pervasive point of discussion at a recent Atlantic pact (NATO) meeting was how to define the word (*Dagens Nyheter*, December 29, 1952). Such early examples show that contemporary struggles with defining infrastructure, exemplified by a 2021 CNN headline "The fight to define infrastructure could change America," are not new.[15] The tension between this plastic, malleable quality, which makes infrastructure negotiable and extensible, and the long-term stability associated with infrastructure is a theme throughout this book, relating to both how infrastructure was constructed historically and how it can be made.

- *Infrastructure is an idea.* The expansion of "infrastructure" relates to societal and political developments aligned to the development of infrastructure as an idea. This development is not just about a term but rather about shifting frameworks and real-world change. For example, higher education (and especially research) in Europe is markedly different after "research infrastructure" was implemented as an administrative, financial, and scientific category, which, in turn, can be related to higher education becoming increasingly global, competitive, and managed by control. Research infrastructure embodies the broad policy shifts toward innovation and usefulness after World War II that consequently aligned with the rise of new public management. Research infrastructure road maps, for example, which impose managerial control at many levels at the same time, become tied to business plans and employ "usage" as a common metric.
- *Infrastructure was not always there.* While many of the ideas, artifacts, and organizational models associated with infrastructure have been around for a long time, we should be careful not to infrastructuralize the past uncritically or to assume that "infrastructure" is just a term without any sociocultural embedding and associations. In fact, infrastructure encodes and articulates a way of thinking about the world. For example, while "research infrastructure" as a framework and idea did not exist in the 1980s, it is now a defining quality of higher education of much of Europe.
- *Infrastructure is inflected.* Any infrastructure is charged or inflected in some way, by standards, values, politics, power, and other factors. I use the term "inflection" to describe the epistemic, diverse, intellectual, social, and potentially radical nature of infrastructure. The infrastructures themselves may not be invisible, but their inflected nature often escapes our attention. We do not necessarily see the racialized or environmental consequences of road systems or search engines,[16] how presentation tools such as PowerPoint structure public discourse and knowledge production,[17] how Zoom as a video platform structures conversation, or how Omeka, a tool for managing and presenting cultural heritage material, privileges particular forms of and perspectives on cultural heritage.[18] The effects of inflections often extend beyond the immediate context of the infrastructure. The freeway system in Los Angeles is not just a neutral transportation infrastructure. It has also been shaped by displacing people of color, segregating people, and creating environmental injustice.[19] The effects are very real on multiple levels. Residents in South LA, which is enclosed by freeways and heavy traffic corridors, are much more likely to have asthma emergency visits or

to die from coronary heart disease than the more affluent, predominantly white residents in the much less polluted Westside.[20]

- *Academic infrastructure is civic infrastructure (and vice versa).* Academic/research infrastructure and civic/societal infrastructure can seem different, but there are many similarities. For instance, the general technological and instrumental framing applies to both infrastructure types. Moreover, higher education is not divorced from society, and research infrastructure has inflection just like civic infrastructure. In this book, I argue that civic and academic infrastructure need to be closely connected analytically and that this connection presents an opportunity both from a societal point of view and for higher education, especially given the transversal and convening role that universities can have in responding to societal challenges and complex problems.

The present study draws on the distinguished history in science and technology studies and infrastructure studies as well as developments in other relevant fields, including digital humanities, disability studies, and environmental humanities.[21] Emergent research that engages with critical and material making is important for the project, including work in feminist infrastructure and visualization, environmental health, critical making, critical and responsible archives, archival records and their imaginaries, intersectional digital humanities, critical design, and designing for difference.[22] The study is also based on practice as well as personal and institutional experience in grappling with infrastructures materially and conceptually as well as in terms of policy. This engaged and experiential component is important to the topic and the narrative.

Unlike many other studies, the emphasis in this book is on how infrastructures have changed and how they can be changed through critical-constructive knowledge and practice. Considerable attention is paid to the *how* of building capacity to mobilize humanistic and humane knowledge to respond collaboratively to problems and challenges. The result is no singular model, solution, or humane infrastructure but rather a humble, hopeful, and incomplete bringing together of experience, frames, and suggestions. This gradual bringing together toward humane infrastructures is evident in the story of the book, hopefully giving the reader—through the book—an understanding of what it is at stake and how to generate institutional and infrastructural change.

Humane Infrastructures is organized into four chapters. Chapter 1, "The Making of Infrastructure," introduces civic and academic infrastructures and investigates the formation of infrastructure as an idea and framework, starting with a retrospective exposition and moving to defining infrastructure as a long-standing problem and research infrastructure as a case study. Special attention is given to the mechanics and discourse surrounding infrastructure work using material from Sweden, the US, Europe, and Australia. The chapter ends with a discussion of the disconnected nature of infrastructure.

Chapter 2, "The Infrastructural Humanities," explores how the humanities engage with infrastructure, particularly how they have negotiated, resisted, and aligned with infrastructural regimes. It also proposes five models of humanistic infrastructure: the solitary scholar model; the cultural heritage model; the digital cultural heritage model; the data model; and the communication model. The final part is a case study of humanities centers and institutes as infrastructure.

Chapter 3, "Instrumentalizing the Humanistic," examines methods, approaches, and frameworks driven by the humanistic and humane that already are or may be recruited to critique, reimagine, and change infrastructures. Such tools include infrastructural inversion and terraforming. The chapter establishes a provisional repertoire of the humanities, which is used to identify some fields with a strong potential for active, critical, and constructive engagement, including urban humanities and disability studies. Individual engagement is traced through a conversation between two scholars on climate change. The issue of instrumentalization is considered as well as a possible toolbox, leading to the capacity building envisioned in chapter 4.

Chapter 4, "Designing Humane Infrastructures," explores the question of how to build capacity to mobilize humanistic and humane knowledge to respond collaboratively to challenges and problems. The chapter opens with a conversation on grounded critical-constructive work and moves on to consider the conditioning of infrastructures, how such conditions build up frames, and how arenas for transversal knowledge production can function as an engine for capacity building. These perspectives are then pulled together in an experimental critique session on an emergent initiative called the "Infrastructure Clinic" (IC).

Why Humane Infrastructures?

Work that seeks to connect ideas, visions, and operational goals with technologies, organizations, people, and communities is infrastructural. Infrastructure is where the rubber meets the road, to use a somewhat antiquated metaphor, and material infrastructures reflect layers of societal, cultural, organizational, and conceptual choices and inflections. The metallic cage built around a hot air vent in a sparsely populated Swedish city during wintertime to make it impossible for an unhoused person to sleep there enacts a worldview and reflects a society.[23] When an Australian policy report states that the story of research infrastructure is one of "the leaps that remake the world,"[24] it also represents a worldview—that research infrastructure will revolutionize science—which renders other perspectives and priorities less potent. Although infrastructures are obviously out there—often for a very long time—they are never *just* out there.

Research infrastructure emerged as a framework during the time we built and expanded HUMlab at Umeå University, an open meeting place for the humanities, culture, and information technology, which I directed for almost fifteen years (2000–2014). While HUMlab is now categorized and presented as research infrastructure,[25] it was not called that for a long time, despite being a technology-rich environment for facilitating research with equipment for visualization, simulation, and cultural databases. Likewise, in 1983, when Larry Smarr and his colleagues submitted a groundbreaking proposal to the National Science Foundation (NSF) for a "Center for Scientific and Engineering Supercomputing," they did not use the term "infrastructure" or "research infrastructure" in their proposal.[26] Research infrastructure (under several rubrics) only started to emerge as a framework in the late 1990s. This logic has major repercussions for how we organize and fund higher education and, indeed, for how we think about the university. My own trajectory, which is reflected in the book, started out from this engagement with academic infrastructure as a scholar and practitioner and as someone interested in policymaking. The research then extended to cover a growing interest in action and change in the academy and outside. A key question became how to build infrastructures that allow such work at scale.

This book was made possible through a large and growing network of interlocutors, including Matt Ratto (who read and reread most of the book),

Francesca Albrezzi (who expertly helped me prepare the manuscript), Lina Lokander (who read the book carefully and helped me remove inconsistencies), Erica Robles-Anderson (who helped me get started and read early chapters), Pelle Snickars (who read some chapters and encouraged me), Michael Goodhart (who helped me get to grips with some things in the text), Anders Ekström (who read parts of the manuscript), as well as Nicole Anand, Anne Balsamo, David M. Berry, Gargi Bhattacharya, Anne Burdick, Jeff Burke, Geoffrey Crossick, Dana Cuff, Cathy Davidson, Gary Dirks, Johanna Drucker, Jennifer Edmond, Sophia Efstathiou, Kate Elswit, Zephyr Frank, David Theo Goldberg, Vendela Grundell, Lisa Yin Han, Margaret Hunt, Jacob von Heland, Theo Henderson, Sara Hendren, Lorna Hughes, Natalie Jerejimenko, Christopher Johanson, Marie Louise Juul Søndergaard, Sally Kitch, Christopher Lee, Cecilia Lindhé, Alan Liu, Shannon Mattern, Tara McPherson, Naomi Murakawa, Jane Nguyen, Darci Niva, Christer Nordlund, Jasmine Nyende, Jane Ohlmeyer, Marisa Parham, Anne Pasek, Miriam Posner, Todd Presner, Rita Raley, Roopika Risam, David Schaberg, Jeffrey Schnapp, Sha Xin Wei, Nishant Shah, Michael Shanks, Eva Stensköld, Sverker Sörlin, Fred Turner, and Michael Watts.

The analyses, approaches, and models presented in this book draw on early work I did in HUMlab in Umeå together with a highly competent and creative group of technicians, faculty, and administrators and through extensive traveling and networking during that time, and they have been developed and accelerated over the past decade in New York City, Los Angeles, and Stockholm. Crucial support has been given from Umeå University, the Wallenberg Foundation, UCLA, City University New York, Balticgruppen, Riksbankens Jubileumsfond, and the Swedish Collegium for Advanced Study. The helpful comments made by two anonymous reviewers of the proposal and first chapter and by three reviewers on the full manuscript have been invaluable. I am grounded in Sweden, Europe, and the US, which means that this is where I found much of the material and context for my analysis, even if I also looked elsewhere. I engage most actively with materials in English and Swedish.

Importantly, the book was also shaped by a series of events that I curated or cocurated since I left Umeå. Between the fall of 2015 and the summer of 2022, I organized or co-organized about twenty events in Stockholm, Uppsala, New York City, London, Los Angeles, and Tempe and online, all of which contributed to this book one way or another, including an important workshop

Figure 0.1
From the "(In)humane Infrastructures of Homelessness" session at UCLA, March 9, 2020. (a) Theo Henderson on the left and (b) the full table of participants on the right. Photo credits: Jane Nguyen (left photo) and John Glover (right photo).

on "Humane Infrastructures" at UCLA on March 9–10, 2020. A group of more than a hundred people gathered in Los Angeles around the following question: How do we create humane civic and academic infrastructures?

Infrastructure means different things to different people. At a session on the "(In)humane Infrastructures of Homelessness," former elementary school teacher and founder of the podcast "We the Unhoused" Theo Henderson was at the center (figure 0.1). At the time of the workshop, he was unhoused. Henderson's podcast brings up issues of relevance to the unhoused community, often through interviews, on-site discussions, and moving about in the city talking to people.[27] He challenges the system and conceptions about homelessness and works to reframe those and make things better for the community. For example, he insists on using the denomination "unhoused" instead of "homeless," and he emphasizes how his community gets dehumanized. His podcast is an important infrastructure for the unhoused community, recorded on a mobile phone with a new episode every second or third week.

The session was early in the program for the workshop and set the tone for the whole event. There was a rich discussion between the participants, also including an architecture professor, a city council member, a political theory and race studies professor, the lead for a nongovernmental organization (NGO) consortium, and several activists.

Henderson reminded participants that, for the unhoused community, infrastructure can be the lack of bathrooms, the impossibility of getting a recognized postal address, or the sweeps that take away their belongings. To Henderson and the unhoused community, infrastructure affects their lived experience and is directly hurtful, reflecting layers of inequity, dehumanization, and lack of access. This is a crucial point and tension. Infrastructure is not just an abstraction, theoretical leverage, or a way to engage with important issues at a distance but is also *lived experience*. In many ways, Henderson was more apt at moving between the concrete and abstract than the academics in the space.

Homelessness is an ongoing crisis in cities such as Los Angeles, where the official pre-COVID-19 count for LA County was about sixty thousand unhoused people, with many more being one or two paychecks away from losing their housing. Homelessness is a deeply infrastructural issue. At heart, homelessness is about access to affordable housing and living space, but the issue reflects layers of inequality and unjust civic structures. For many Angelinos, homelessness is materialized through the unhoused people they see in "their" neighborhoods and through objects deemed undesirable for the areas in question, such as tents, bags, and "stuff" on sidewalks, under bridges, in cars or RVs, and at other locations where some provisions for unhoused living exist. This informal infrastructure, largely a result of the uneven distribution of the benefits of formal infrastructures, gets regulated and enforced by city workers.[28] While the property of housed citizens is fiercely protected, the belongings of unhoused people are literally trashed in regular sweeps—not stored away but thrown away.

> **This is for Joe whose heart medication was thrown out in a sweep, Chris whose mother's ashes was destroyed in a sweep, Khan who begged to keep his bedding, Janet Garcia who lost her cleaning supplies for work, including a cherished vacuum cleaner she named "baby shark."**[29]

Unhoused people are dehumanized by such treatment and many other similar societal and community actions, ranging from comments on community Facebook groups to closing public restrooms at night to stopping unhoused people from charging their phones at coffee shops to creating and enforcing various regulations meant to criminalize and remove unhoused people to outright physical attacks. So-called NIMBYs (people who subscribe to a "not in my backyard" mentality) organized to oppose plans to build shelters and even to house sick people in a hotel during the COVID-19 pandemic.[30]

It is easy to take a stand from a distance—for example, against NIMBYism—but things do not normally start with organized actions or attacks. Rather, they often come from feeling uncomfortable and being worried about your own property and "your" pavement, alleged crimes and drugs, and the safety of your children. These worries easily and problematically get attached to the unhoused community and to specific individuals. I lived in a well-to-do community in LA with a small, unhoused population, but I sometimes had to fight with myself when I felt that "my" space was taken up by people who sometimes seemed strange or hostile or who were just in my way.

It is important to reflect on one's own reactions, including their rationality and humaneness. It also helps to treat people as humans and talk with them. At the same time, there is no getting around the fact that there are layers of challenges at play, including access to mental health care, drug abuse, lack of affordable housing, inadequate coordination of resources, and systemic inequity. I could see a negotiation going on in my own community and got increasingly worried about the tone and the undercurrents. I was especially troubled by the generalized and, at times, racialized reaction to the unhoused community and to visitors to the community. I felt the need to react, and however small an intervention, I did speak up a few times, including at an important community meeting.

One problem is that worries and fears (and sometimes malice) give rise to very unforgiving and inhumane treatment. The "solution" must be structural, systemic, and humane, but that does not give us a license to be inhumane along the way. At the "Humane Infrastructures" event, Jane Nguyen, one of the leaders for the organization Ktown for All, told us how her work as a neighborhood activist started as a reaction to the resistance to build a homeless shelter in Koreatown in LA in 2018.[31] The philosophy of Ktown for All is simple and humane but not as easy as it should be: care about your neighbors!

The inhumanity I encountered when preparing the session was of a kind I did not know much about and to an extent that I did not realize beforehand, although homelessness is a defining quality of Los Angeles. Entrenched and systemic, it is part of a repeating, historical pattern. At the same time, during this research, I met some of the most humane people and initiatives I have ever come across—people trying to make things better by building platforms, gathering resources, organizing communities, and working together.

Louis Tse, executive director of Students4Students, mentioned in passing at the "Humane Infrastructures" workshop that as a student at UCLA, he slept in his car in order to fund the student shelter he cofounded. Much like Jane Nguyen, he reacted to people around him being in need. Importantly, their resistance and that of the unhoused community, activists, progressive city council members, organizations, and fellow citizens is also part of the infrastructure of homelessness.[32]

Louis Tse's and Jane Nguyen's work makes clear the connection between civic and academic infrastructures. Not only is the homelessness crisis in Los Angeles a responsibility for UCLA through the institution's placement and role as a public university, but homelessness is also part of the inner life of the institution. Estimates show that 5 percent of University of California students are homeless.[33] This is not something that is immediately visible on the lush UCLA campus or in how the institution frames itself. In the UCLA 100 centennial campaign (concluded in 2019), for example, homelessness was a theme listed under impact stories and themes, but the issue is clearly placed elsewhere. The UCLA website states that in order to confront the homelessness crises, "UCLA moves off campus and into the streets" to bring "knowledge to the streets."[34] This is a clear example of placing the challenge elsewhere, in the city, rather than also acknowledging the housing insecurity that many on campus face. Such distancing is also demonstrated in the lack of data collection on homelessness among students, which can be regarded as an infrastructural question.[35] There is something deeply troubling about this displacement, especially given that UCLA is so rich in resources, knowledge, and engaged faculty and students.[36]

This separation is common in academic scholarship. We tend to look elsewhere rather than at ourselves, which also makes it easier to keep a critical distance from the objects we study. For example, we may analyze the power structures associated with space and architecture in society but pay little attention to how our own institutional space is organized and inflected.[37] For instance, what kind of office space is given to staff and why? If we actually do pay attention, the next step must be not only to question our own arrangements and organizations—which requires us to acknowledge our own complicity in such systems—but also to consider how they can be changed to reflect the values and principles that are important to us. Such active, critically constructive engagement is necessary in relation to civic as well as academic infrastructures.

The privilege of the "Humane Infrastructures" event was palpable. UCLA is one of the best and well-resourced public universities in the world. The event took place centrally on campus in the single-story, ranch-style building that houses the faculty center—a place that exudes networking and power. We were in the California Room, which is an expansive, grand room with chandeliers, a fireplace, and extensive patio. We chose this venue because it was an available signature space that was spacious, had flat seating, and was capable of accommodating round tables. We did not want a theater or lecture-style space. Rather, we wanted a space where discussion would occur around and between round tables rather than upfront on a stage. It did not hurt us to use such a premium space for the workshop, but it also felt slightly uncomfortable.

We wanted the infrastructure of the event to manifest the values and topics that were at the center of the conversation, and we tried to call attention to some of the tensions at play, including the privilege of UCLA as an institution. There was a reason why we started the workshop—following the introductory conversation—with the session on homelessness. We included a land acknowledgment on our website and expressed it aloud before our proceedings. We also addressed different levels of access and provided accommodations to be as inclusive as possible. Several participants emphasized the need for reflexivity, and there was something about the setting, timing, and format that made such perspectives very real.

The COVID-19 pandemic was imminent, and in the afternoon of the second day, there was a message that campus teaching would be suspended the following day. We had hired several architecture students to assist in facilitating the workshop. They did so expertly for as long as they could, but due to the rising circumstances, they were forced to leave early to meet up with their departments before the shutdown, which made the gravity of the ongoing developments very clear. In addition, the cost-of-living adjustment (COLA) protests and strikes at University of California campuses, starting with UC Santa Cruz, were underway.

Several participants asked us whether there would be a strike at UCLA, in which case they would not come for the event. While there was no COLA strike at the time at UCLA, the movement was clearly present through the days. For example, after being granted permission, I read aloud the rationale given by one of the invited participants, Nick Mitchell at UC Santa Cruz, for not coming:

> **After four out of my five teaching assistants were terminated on the Friday before last, I composed messages to inform all of those to whom I had professional obligations in the next month and a half to inform them that I would need to cancel so as to redirect my attention to the defense of our students. . . . At this point we are fundraising to try and ensure that the ten international and undocumented students who have had their spring quarter employment terminated are not deported as a result of the university's response to them.**[38]

As our conversations over the two days emphasized the importance of being reflexive, looking at ourselves as well as others, there was a collective acknowledgment that a path from such reflexivity to action was also needed. One of the richest discussions was during a session on strategies for such critically informed action with participants with a strong—or even career-defining—interest in developing such frameworks. The prompt for the conversation was the invitation to imagine or propose a critical-constructive toolbox that could be used across individual methods and approaches.

The design processes advocated by scholars such as critical maker Matt Ratto, crip technoscientist Aime Hamraie, and critical-feminist designer of intimate technologies Marie Louise Juul employ critical categories not just as perspectives but also as a central design premise and a means to inform action. Artist and scholar Romi Ron Morrison's work also challenges conventional design logic, including the embedded assumptions of sociotechnical systems that aim to calculate Black life, and their artworks investigate how boundaries, fissures, and inconstancies can be part of practicing new spatial relationships and epistemologies.[39] Ursula Heise, English, talked about terraforming as an imaginary and constructive process in relation to her work on science fiction and environmental humanities, where basic infrastructure and ecology get reimagined and reshaped. As an engineer, researcher, and artist, Natalie Jeremijenko stressed during the conversation the role of imagination and humanistic ability to reshape and rework conceptual frameworks, which she does herself through her Environmental Health Clinic and mutualistic systems design. One thing that struck me about this conversation was the careful attention to the inflected nature of current conditions as well as change itself and the complexity and necessity of critically and humanely driven processes of change.

While it did not have its own session, "scale" was a related issue that connected to the toolbox discussion and resonated through most of the other program slots. UCLA librarian Zoe Borovsky talked about the tension between the large, generic infrastructural systems mandated by the library and the practice of carrying out small-scale projects close to specific research questions and materials. It is not that the small-scale projects are not informed by standards and existing software but rather that such a "high-touch" approach is largely contradictory to the ideology and resources associated with infrastructure work. A key challenge is to scale up successful, small-scale models and processes, which is a question not necessarily of technology but rather of staffing and cost distribution. Urban planning professor and architect Kian Goh made a similar case for "big justice."[40] The focus here was not moving beyond the particular and small scale but rather expanding the scale from the local to the more than local in order to make change—for example, when striving toward environmental justice in urban planning.

The discussion of scale also connected directly to the overall framing of the event: How can we build humane- and humanities-driven capacity to respond collaboratively to societal challenges and complex problems? Building such capacity—infrastructure—requires extending from the particular and small scale to the more generic and large scale and engaging across domains, competences, epistemic traditions, methods, and institutions. There is something distinctly un-humanistic about the idea of scaling up, including the risk of erasing the particularity of situated experience, local culture, rich context, and disciplinary expertise. "Scaling up" is also an expansionist notion part of management or business language, frequently associated with a corporate, institutional, instrumental, and neoliberal agenda, which can understandably be provocative for the humanities.

Building capacity across domains is difficult, particularly on one's own terms, which is why such operations tend to be precarious. At the workshop, Matt Ratto—who has built extensive programmatic structures for critical making—argued that it does not seem right to "always stay on the fringe." This is why it is not enough to build innovative infrastructure. We also need to change the logic of infrastructure building to change long-term opportunities and impact. Being on the fringe is important, but so is institutional stability and long-term care. Digital humanist Roopika Risam pointed to how we need to significantly change our infrastructures or calls for generosity and care will be meaningless.[41] Care must be part of the systems that we build, including a fair distribution of resources, co-credit systems, accommodation of multiple perspectives, collaborations that benefit others as well as ourselves, concern for the environment, and a just, actionable "logic." Such systems must be simultaneously responsive and responsible.

The Humanities as Anchor Point

I argue that the humanities can contribute to such logic-shifting work, not by themselves but with others, and hence be part of the building of humane infrastructures. This is not an institutional or territorial argument but rather one based on engaging the critical, challenging, and curating powers of the humanities to collaboratively create new knowledge, facilitate conversation, and "make things happen." This vision is predicated on working respectfully and generously with humanistic scholars, other fields, external partners, communities, and stakeholders. In what follows, I will introduce the humanities briefly and attempt to trace the humanistic powers mentioned above.

"The humanities" is an intriguing label in that many who are subsumed under conventionally humanistic disciplines such as English, history, linguistics, and philosophy do not seem to have a strong connection to it, albeit being enfolded administratively. This weak sense of belonging on organizational grounds probably constitutes the most straightforward definition of the humanities: simply a list of disciplines that have emerged over time under this label, with their own epistemic traditions.[42] This definition may not be very satisfactory but would get around the apparent difficulty of establishing a unified set of objects, methods, purpose, or politics for the humanities beyond "the human condition," studying humans in their cultural context, or studying the past and present meaning-making practices of human culture.[43]

For others, however, the term is deeply meaningful, and there is a long and extensive history of engagement in and for the humanities, often fighting for their survival, proclaiming their value, or projecting their role for hopeful futures. This hope, frequently driven by the acuity of societal challenges, can be found in a whole range of "new humanities" such as environmental and urban humanities (see chapter 3). Here, the humanities are centered, often to indicate a humanistic perspective on a broader theme (e.g., the environment, urbanity), but the humanities imagined are generally different from the current humanities, which is why structural or infrastructural change is advocated, sometimes in a supercharged manner. Others—and also some of the same hopeful people—take part in researching the humanities as part of science and technology studies, university studies, history of humanities, or history of knowledge. Such analytical charting can, for example, be disciplinarily grounded, cover the recent history (e.g., postwar humanities), go back several centuries BCE, or be based on frameworks such as "circulation of knowledge" or a critical taxonomy of justifications for the humanities.[44] Unsurprisingly, there is often a tendency to remain with the disciplinary core rather than engage with (newer) liminal areas such as queer studies and digital humanities. Although such historiographic and critical work on the humanities naturally imposes interpretative frameworks, it is generally relatively distanced. Other scholarship is much more directional and programmatic—for example, arguing that we need to move away from the anthropocentric orientation of the humanities.[45]

There is no getting around the fact that categories such as "humanities" are naturally exclusionary. The humanities and humanistic are central to the argument of this book, but they are seen as an anchor point, arena, or a key rather than a disciplinary or institutional container—not primarily by themselves but together with other areas and the outside world. Among other things, the humanities offer "critical thinking" and deconstructive capability to such collaborative contexts, not uniquely or unequivocally but strongly, almost constitutionally.[46] Humanities approaches are often taken to be interpretative, holistic, and non-reductive, which can be contrasted with more pragmatic fields.[47] Another way of framing this critical power is to assert that the humanities can be used to "think with."[48] Indeed, humanists can be facilitators of meaningful dialogue and activities across traditions and perspectives—deploying a kind of "curating power" to get actors, ideas, and perspectives together and collectively make things happen, including imagining, changing, and building infrastructure. They can also participate as scholars with disciplinary knowledge. Disciplinary, domain-specific knowledge from the humanities relevant for the "climate" (to pick a topic) might, for example, relate to the study of historical energy transitions, cultural and linguistic taxonomies of ice, the ideational construction of "carbon footprint" in the media, the literary or cinematic representation of catastrophes, the history of scientific visualization, extractive mediation and the seafloor, or the anatomy of environmental racism.[49] The resultant conversations and activities, necessarily critical and constructive, must build on curiosity and engagement and be contingent on inviting others within and outside the university, working with them, and being willing to change and be changed. Participants may identify with the humanities or not; the important thing is that there is mutual respect, openness, and a deliberated force forward.[50] The invitation expressed here is part of the direction presented in this book, but it also applies to the book itself; it is meant to be open and curious, interested in different knowledge traditions and backgrounds, and in search for dialogue.

For the purposes of this book, the humanities are taken to constitute a broad category that also includes—or, perhaps better, interfaces with—many areas and initiatives on the boundary or just outside the institutional humanities, including disability studies, information studies, design, science and technology studies, queer studies, and the arts. Along these lines, the adjectival "humanistic," often a more useful term, is employed to refer to people, perspectives, and thinking that can be associated with the critical, cultural, or historical perspectives that are often (but not always) attributed to the humanities. There is no claim, however, that the academic humanities have a monopoly on certain perspectives or types of knowledge, but rather that there is a wealth of concentrated and aggregated domain-specific knowledge, thematic orientations, and general approaches in the humanities that are underutilized in the academy and can help ground outward-facing engagement together with others to support structural change within and outside the university.

Another reason why the humanities (and the arts) are relevant for the current project is their potential contrariness. They can be oppositional, edgy, sharp, and not tame, even dangerous. This criticality and oppositionality are more constitutional—closer to the skin—for the humanities than for other fields, which can be a problem, in fact hindering engagement or development, but is also clearly an asset.[51] For example, the ability to resist notions such as "wicked problems" or "planetary health" is vital, but where suitable, it is also useful to engage with these and other concepts and frames constructively—critiquing, contextualizing, and pushing back, but also querying, shifting, negotiating, changing, reframing, suggesting alternatives, designing, and so on. It is not that humanists are not stuck in different schools of thoughts, paradigms, and epistemic commitments, but they can nevertheless have a more general ability to challenge worldviews and take systems apart. This "challenging power" and ability to make people think differently about the world can be very powerful, especially if combined with constructive, imaginary work.

1

The Making of Infrastructure

Definitions typically describe the foundational, necessary, and essential qualities of infrastructure—for example, the "underlying foundation or basic framework (as of a system or organization),"[1] "a collective term for the subordinate parts of an undertaking; substructure, foundation,"[2] "sets of collective equipment necessary to human activities,"[3] the "physical components of interrelated systems providing commodities and services essential to enable, sustain, or enhance societal living conditions,"[4] or the "undergirding of modern societies."[5] These interpretations construe infrastructure as the key structure on which other things are built and made possible, the contours of which are material, technological, and systemic. Infrastructure has been constructed—templated—over time in this vein, although it is simultaneously a flexible and expansive concept.[6]

Both of these tendencies can be seen in land defender and Indigenous environmental knowledges scholar Anne Spice's instructive discussion of "invasive infrastructures," where she recounts a dialogue with Unist'ot'en Hereditary Spokesperson Freda Huson in the face of pipeline projects threatening their territory. Here, what is conventionally framed as "critical infrastructure"—oil pipelines—gets reclassified as invasive, and the term is used to describe the clan's best berry patches:

> **In Unist'ot'en territory in northern British Columbia, Canada, clan members of the Wet'suwet'en people have built a permanent encampment in the pathway of numerous potential and proposed pipelines. In response to the characterization of these pipeline projects as "critical infrastructure," the camp's spokesperson, Freda Huson, notes that the pipelines were proposed to run through the clan's best berry patches. By resisting pipeline construction, she explains, "what we're doing here is protecting our critical infrastructure."[7]**

The concept "critical infrastructure," part of systems vulnerability (or vital systems security), is thus challenged and appropriated, and in doing so, other facets than extractive technologies and legacy infrastructure are centered, at least discursively for the moment.[8] Although berry patches could conceivably be included among "processes, systems, facilities, technologies, networks, assets and services essential to the health, safety, security or economic well-being of Canadians and the effective functioning of government," it does not seem likely that they will be prioritized as a critical infrastructure of national import.[9] What is foundational (and critical) is a matter of policy, power, and convention.

Despite many interdependencies and overlaps—including a sense of complexity and heterogeneity—it can be hypothesized that the foundational quality attributed to "infrastructure" is one of the basic criteria that separates it from related terms, such as "network," "system," "technology," and "logistics." It should also be noted that infrastructure has an affinity with networks as well as with distribution and circulation. As an aggregate term, infrastructure becomes the "built networks that facilitate the flow of goods, people, or ideas and allow for their exchange over space"[10] or the "socio-technical apparatuses and material artifacts that structure, enable and govern circulation."[11] As we will see in chapter 2, "circulation" is one of the frames used to model humanistic research infrastructure. Another term, "platform," shares a sense of being foundational or supportive but usually references entities that are smaller, less generic, and more likely to be corporate, although the boundary can be fuzzy and dynamic.[12] Difference in complexity can also help us place "system" in this semantic field. Literary scholar Clifford Siskin argues that while "system" makes up a central component of infrastructures, it is only one of the many units that constitute them.[13]

This chapter invites dialogue with fields such as infrastructure studies, design, and science and technology studies. It is intended to inform the subsequent discussion of how infrastructure can constitute a conditioning framework (chapter 2) and how infrastructures can be modeled and shifted, not just concrete infrastructures but the ideas and frameworks associated with infrastructure (chapters 3 and 4). For this purpose, I engage the concept of "infrastructure" historically, creatively, and critically through assembling a range of resources and perspectives, drawing on my background as a linguist and digital humanist.

Tackling Infrastructure

This section outlines a journey trying to tackle "infrastructure," in a sense documenting some of what was learned writing this book in a rather long and collaborative back-and-forth process. What is the role and contribution of infrastructure to the story presented? While the trajectory described is iterative, it extends across several key axes: from research infrastructure to civic infrastructure, from actual infrastructures to infrastructural frameworks, from epistemological to societal engagement, and from infrastructure to capacity.

Looking at Infrastructure

One thing that makes infrastructure intriguing and alluring as a concept is that it simultaneously calls forth abstract aggregated systems and material, often technological, artifacts. According to media theorist Brian Larkin, there is a duality between infrastructures as objects that are the foundation on which other objects work and their capacity then to operate as systems.[14] The road and traffic lights I saw outside the window of my coffee shop writing this text fragment (figure 1.1) are part of the massive road infrastructure of Los Angeles—which, in turn, is part of a national infrastructure—but this specific stretch of road and the local traffic signals would not seem to constitute infrastructure by themselves, and neither would the individual cars and drivers.

The infrastructural quality of this scene is arguably contingent on the system, the road continuing, the traffic signals being part of a system, and

Figure 1.1
Los Angeles local traffic.

so on. In Los Angeles, there are 7,500 miles of city streets and a signal surveillance and control system with 4,850 signals.[15] A computational "driver," as in autonomous cars and robotaxis, appears more infrastructural than a human one not only because it is technological but also because it is clearly part of a networked system.[16] It is probably not surprising that a Google image search for "infrastructure," carried out at an early stage of writing this book, resulted in photos of complex interstate intersections and heavily trafficked road systems rather than photos of individual roads or cars.[17] The Google images—an infrastructural representation in their own right—also included many photos where the systemic and networked materiality of transportation infrastructure is signaled through visualizations of movement and flows, in particular by means of blurry long-exposure photos.[18]

The picture that gradually came into view positioned infrastructure as much as an idea as a set of material and conceptual configurations. This idea or frame—in its complexity—has emerged over time in relation to a set of societal, cultural, and technological developments, including the introduction of railroads, the postwar rebuilding of Europe, deregulation of public works, increased securitization, the maintenance and reimagination of colonial infrastructure in the Global South, globalization of higher education, and large-scale digitization and growth of "data."[19] It is not that the world did not have roads, ports, electricity, or data before the concept of infrastructure emerged more comprehensively toward the latter half of the twentieth century but rather that "infrastructure" has become a pervasive frame for talking and thinking about the world and for building that world. A parallel can be drawn to the emergence of "the environment" as a category and idea after World War II, which gave us a way of talking about and understanding an interconnected and vulnerable natural world across scales in a way that was not possible before.[20]

Early encounters with "research infrastructure" in the context of research funding and humanities adaptations of this framework—essentially where this story began—made infrastructure seem like a set, closed category. More generally, infrastructure often appears fixed and difficult to change, at times unchallengeable. Imagine changing a national road system, global shipping logistics, a university reimbursement system, or how a city thinks about urban development. Where does one even begin? One important reason for this fixedness is the emphasis on generic and standardized solutions, which makes it difficult to imagine and implement local, non-standardized infrastructure.[21] Also, infrastructure studies point to how infrastructure gets naturalized and is sociotechnologically embedded.[22] Hacking a local traffic light or repainting the lines outside your home to change traffic patterns will likely not work, especially not in the long term. Even authority-driven challenges, such as the so-called road diet implemented to improve conditions for bicyclists and pedestrians in my LA neighborhood, can be resisted—in fact, this was resisted and subsequently reverted.[23] Sociotechnologically embedded expectations—such as the centrality of the automobile in LA transportation infrastructure and urban life—contribute to shaping infrastructure, just like rules, procurement procedures, and epistemic and professional traditions.

Questioning Research Infrastructure

If one expectation about research infrastructure came across when the term first started to appear in discussions in HUMlab, it was that a cultural humanistic technology lab did not seem part of the definition. "Research infrastructure" had developed since the early to mid 2000s In Sweden to denote, prescribe, and organize large-scale technological instruments and facilities for mainly science and technology research. We came up against a rigid framework that not only appeared to exclude us but where possible paths forward seemed limiting, essentially suggesting a standardizing and instrumental logic dependent on tools and databases. We realized that we needed to become more systemic—infrastructural—in our own thinking and to consider existing conditions critically as well as the possibilities of humanities-driven infrastructures beyond the template. One step was to move from individual cases (projects, tools, environments) and internalized practices (accumulated lab-based knowledge, knowhow, experience) to connecting conceptual and material aspects of infrastructures in a more structural manner.

Design principles offered a way to bridge the ideational level and material infrastructure "without getting caught up only in detailed infrastructure and/or the abstract visions typical of the discourse on cyberinfrastructure."[24] When designing a humanistic lab space (which we were doing), for instance, relevant design principles might be "translucence" (the ability to see and hear "through," to be separate as well as part of something bigger) and "memory/context" (traces from previous projects, people, and experiences interwoven into the space in different ways).[25] Such design principles connect to the overall vision—in this case, creating a meeting place for the humanities, culture, and information technology—and the actual implementation, including not

Figure 1.2
A nook in the HUMlab space, 2012. Stephanie Dean in the foreground.

closing off sections totally, creating nooks (as in figure 1.2), letting projects live in the space, and having room for past prototypes.

Johanna Drucker and I described intellectual middleware as "the ways tools structure our arguments or express thinking in protocols programmed into these platforms."[26] The notion of middleware allowed us to look systematically at the belief systems articulated through different types of platforms—for example, how database tools for cultural heritage contain and express particular ways of thinking about such material (including "the collection") at every level of production. We maintained that such argument structures or protocols are often not explicitly acknowledged and need to be attended to when planning and building tools and platforms. The concepts of design principles and intellectual middleware suggested ways of articulating and framing a shift from research infrastructure as "pure" high-level instrumentality to having a stronger epistemological and ideational grounding in terms of both policy and material infrastructure.

When looking more closely at research infrastructure as a framework, another type of middleware emerged as part of the underlying logic (principally a science and technology-heavy model of research and research support)—namely, a set of mechanics used to assert or co-evolve that logic, essentially conditioning how research infrastructure can be imagined and made. As we will see in this chapter, research infrastructure as a model has been systematized and standardized through powerful mechanics, including funding priorities, definitions, road maps, and organizational logic.

The Shaping of Research Infrastructure

The interplay between such frameworks, underlying ideas, and actual infrastructure—infrastructuring—takes place here and now but is also a reflection of how research infrastructure has been shaped over time.[27] Some of this history, which ranges back to the 1990s, is traced in this chapter. It became clear that this is not just a linear development but also a clustering of ways of thinking about research infrastructure that emerge as a result of negotiation, often involving previous technological and institutional layers. We noticed how certain ways of thinking about the humanities and their relation to infrastructure (equipment, tools, materials, etc.) were embedded in infrastructure policy and visions, forming models of humanistic infrastructure.

When we met with humanities deans in Sweden at their annual conference in 2018, it became clear that research infrastructure was something rather unfamiliar associated with a technological imperative. However, when the discussion moved to their work, challenges, and visions, it also became apparent how infrastructure could be potentially recruited as a tool and could help to consider how such ideas can be manifested.[28] Infrastructure was a conditioning template as well as a vehicle for imagination and change. In either case, it appears that infrastructure can carry—and be infused with—ideas, values, direction.

What ideas, values, directions? It turned out, for example, that European research infrastructure was expected to contribute to "European cohesion."[29] Calls for funding proposals clearly targeted science and technology through their language and framing—a conditioning that recurred in different modalities, including images and leaflets. An EU infographic states that "Research infrastructures make science happen" and the associated visuals portrayed a traditional lab setup with people in white coats, miscellaneous lab equipment, chemical formulas, and reference to computing systems, scientific data, and sharing scientific discoveries.[30]

In such discourse, the inflection was typically not announced but rather naturalized or suppressed. Definitions were often given by example, taken for granted. In the infographic, the examples given are from medicine and science, and the language and framing (e.g., "scientific discoveries") exclude other areas. The transformative revolutionary sentiment frequently attached to infrastructure policy also comes across as naturalized. For example, the Swedish government sees investments in research infrastructure as a way to be (or become) a leading research nation, supporting breakthroughs and stimulating innovations.[31] Large infrastructural projects such as the European Spallation Source (ESS) are fueled by such faraway hope, sometimes leading to a frenzy as when a Swedish state agency tweeted about Nobel Prizes on the horizon during an ESS conference.[32]

It is not that the imaginary stance is wrong per se but rather that there is often a disconnect between long-term visionary speak and the actual making and function of infrastructural systems. Even when the goal seems "good" and worthwhile, there is a risk that hopeful visions get divorced from grounded interactions in and with the world, hence limiting choices and possible pathways.[33] This disconnect, which is also powered by the mechanics of infrastructure, can be an obstacle to changing infrastructure as both a frame and implementation.

The World Keeps Creeping In

Much of the early-phase work centered on the academy, but gradually the political and sociocultural aspects of research infrastructure became foregrounded. The world kept creeping into research infrastructure, which made it more difficult to keep the two apart analytically. Research infrastructures, even when based on non-radical epistemologies, are about not only knowledge production but also politics, power, and people. For example, they connect to the dominant narrative of civic infrastructure, which has been one of progress and improvement. Infrastructure is taken to serve society, make things better, and create a better future.

Even when we talk about "crumbling infrastructure," this basic frame is not necessarily challenged, as there is a sense that infrastructure can be restored or repaired to some ideal "good" function. This sense of reparability, however, has been usefully challenged through scholarship that emphasizes the role of maintenance, repair, and labor in the making of infrastructure.[34] For instance, as environmental anthropologist Nikhil Anand shows in his research on water infrastructure in Mumbai, maintenance work is much more than predictable processes of restoring relations or original instrumentality. Rather, such work is often characterized by innovation and creativity.[35] In his case study, the system as a whole is conditioned by substantial multiple-tier water leakage that serves as a kind of counter logic to the idea of precise measurements and audit control.[36] Although presumably much more "under control" than failing civic infrastructure, research infrastructure is also continuously made, maintained, and changed with and through people.[37] It is negotiated and packaged and often more open-ended than it may seem.

It became increasingly clear that although infrastructure is useful in many ways, it is not benevolent or evenly distributed. In the world, infrastructure has certainly contributed to public health, services, and economic growth in many places, but it has "equally been shaped by, and contributed to, widespread inequality and unsustainability."[38] Infrastructure building is so intimately connected to resource use, extraction, and exploitation that nature has been called the "ultimate infrastructure."[39]

While the footprint and impact of research infrastructure are more limited, it imposes conditions, priorities, and restricted access in a managerial frame that seems more exclusionary than inclusive. Research infrastructure has considerable environmental and extractive impact through facilities such as telescopes, use of computational and cloud resources, and everyday operations.[40] Even though there are policy-level directives to decrease the environmental impact of higher education in a country such as Sweden, infrastructure is rarely singled out. In fact, the word "infrastructure" is not mentioned in the Guidelines for the Climate Framework established by thirty-seven institutions of higher education in Sweden in 2019.[41] There are other developments, however, that more directly oppose the heaviness and expense associated with infrastructural templates—for example, initiatives on low-carbon research and minimal computing.[42]

There is a technological sheen to infrastructure that contributes to keeping it apart from the world it affects—for example, through environmental harm. As historian Rosalind Williams notes, processes, power relations, and people are concealed through the autonomy given to things.[43] We discovered that there did not seem to be place for categories such as "gender," "power," and "environment" in the research infrastructure policy reports investigated, even in reports on humanistic infrastructure (interlude, chapter 1). This analytical inattention, which mirrored the lack of attention to epistemological inflection in many of the same reports, seemed like an obstacle when considering how to engage infrastructure as a frame constructively.

Such critical reflexivity also needs to extend to the use of labels, whether "societal challenges" or "research infrastructure." I was once asked by a European center for digital humanities whether I thought changing their official name from "center" to "research infrastructure" was a good idea. Not everything is in a name, but I argued that the choice was contingent on where they wanted to go, but because "infrastructure" has this technological, instrumental, and generic reading, it would require a charge or intervention of some sort to move beyond the idea of a support center or being a node in an instrumental, national, and pan-European logic already in place. For me, this sensibility was attached to a growing awareness of the humanities as a place that could and should challenge and push epistemological and sociocultural frames—many infrastructurally embedded—within and, increasingly, outside the university.

Charging Infrastructure

This shift in attention toward the world was not immediate. One precursor in HUMlab was a set of collaborative outward-facing initiatives that challenged established frames one way or another. Examples included working with a municipality organization to empower school children to ideate narratives that were performed as a blog opera on the main stage of the regional opera house, using a faceted browsing interface instead of regular search functions to explore European cultural heritage materials, or talking to a bank about where they might be going in an increasingly digital world (space, physical money, data mining, etc.). In these cases, humanistic questions and knowledge provided a foundation. For example, representatives from the bank were invited to a dialogue with a media historian who had studied multimedia use in American mega churches (including the digitization of bibles), a Moscow-educated architect interested in digital representations of spatial flows, and an ethnologist who had analyzed the gendered nature of industrial spaces and machinery in the North of Sweden (and had spent a couple of days in a local bank).[44]

In the lab work described above, technology functioned as a form of boundary object, bringing people and ideas together but also enabling us to critique, challenge, and shift sociotechnical systems.[45] Importantly, technology was not just a prop or a rationale but rather something we engaged constructively: we built and prototyped things, mostly in an experimental and conceptual fashion. Like technology, "infrastructure" has this boundary capacity, but in contrast, it seemed to lend itself more easily to shaping the conditions of academic work at a systemic level. As the focus of the project shifted toward the world, infrastructure also became a way to engage with the larger sociotechnical contexts in that world. The concept of infrastructure provided a grip, not because it came with a template but rather because that template could be potentially challenged and shifted, ideally offering an opportunity to reimagine and change our lifeworld.

An example is provided by anthropologist Laura Bear when she asks whether infrastructure can make humans more visible, extend political community, and be generative of care.[46] Bear explores austerity policy historically and contemporarily through the Hooghly River as an important artery for trade between India and South Asia, Europe, and East Asia. Her investigation engages with infrastructure as a place for investment and capital growth (rather than public works and non-precarious labor) where austerity and financialized sovereign debt serve as a principal mechanism. In providing suggestions—including a social calculus, new structures of financing, mutuality as a key principle, and new international organizations—she brings a strong constructive element to her critical analysis.[47] Bear's critique also operates on the level of concepts and frames. She argues that we should question what infrastructures are for and consider using other names instead, such as "public works" or "citizens' commons for the public good."[48] A similar call for conceptual directionality can be found in urban planner Hannah Wright's argument for keeping "green infrastructure" ambiguous to embrace both environmental and socioeconomic interpretations, combining theory, policy, and practice.[49]

This active infrastructural direction was connected to a growing emphasis on the role of humanistic knowledge in responding to so-called grand or societal challenges, wicked or complex problems, and missions, all admittedly complicated and convoluted concepts.[50] For example, societal challenges were made central in the European Union in the Eighth Framework Program (FP8, 2014–2020) after becoming increasingly important in FP7 (2007–2013).[51] For us, infrastructure as a concept and idea seemed to offer a point of insertion for carrying out critiquing, imagining, and rebuilding work that could affect the conditioning and framing of our lived world as well as its organizational and material buildup.

Beyond the Humanities

While the notion of "humanistic infrastructure" functioned well for considering conditions of knowledge production within the humanities and for bringing in critical humanistic perspectives into infrastructure work, it increasingly seemed advantageous to move beyond the institutional humanities, especially with the outward-facing direction that continued to develop collaboratively through a series of events, projects, and emerging perspectives. Directionality became stronger through emphasizing how infrastructures could become more meaningful, caring, compassionate, critical, revolutionary, efficient, and life affirming, making "humane infrastructures" not only an ideal goal but also a useful boundary object—something to come together around.

The infrastructure needed for that coming together would likely not be a new fancy building on some campus with a donor's name on it but rather a dynamic multisite, distributed, networked, and synergetic model, pulling together resources broadly in relation to specific themes, projects, and conversations. Such a "contact zone," to borrow Mary Louise Pratt's term, would have to be a humane infrastructure in itself—generous, creative, curious, open, productive, unafraid—but it would also have to strive to create better and more humane infrastructures in the world and the academy.[52] The idea was not for the humanistic to miraculously generate solutions to the world's problems but rather to facilitate dialogue, bring diverse competencies together, and work with different communities and their infrastructures. The humanities, broadly speaking, were seen as an underutilized resource that could and needed to be activated more strongly, not as an add-on, alibi, perspective, or a solution but rather as a full-stack collaborator.

One challenge for imagining this type of capability was to consider how humanistic and critical knowledge (perspectives, theories, methods, etc.) could be packaged sensibly and effectively beyond specific disciplines and themes. This became a recurring topic for activities organized around humane infrastructures, and some examples discussed were infrastructural inversion, crip technoscience, and mutualistic systems design.[53] The issue of responsible instrumentalization is central and complex, especially as the very idea rightfully might seem contradictory to the criticality of the humanities. A key realization, however, was that packaging "content" can be an imaginary, creative process, the result of which can help break out of existing infrastructural frames.

The University as Capacity

A concern that emerged in the context of putting things together (packaging) was that despite the capaciousness of "infrastructure," its range was limited. This concern had to do with the importance of building networks, knowledge, and resources broadly and being able to respond responsibly to a variety of challenges, themes, and problems—that which pulled together resources dynamically and made them decipherable and accessible. Again, not everything is in a word or a concept, but a notion that kept coming up was "capacity." In some ways, it is as laden as "infrastructure" if not more, and for example, it suggests a volumetric logic.[54] At the same time, as argued in chapter 4, capacity can be nimbler, more aggregate, and, in some ways, even more generic than infrastructure. Humane infrastructures were still the ideal result and what we need to strive toward every day, whereas capacity became what we need to build to improve our chances for systemic change—more specifically, capacity to mobilize humanistic and humane knowledge to respond collaboratively to challenges and problems (and other "things").

Such capacity can exist in many forms and institutional contexts, but for the purposes of this book, the academy provided one important starting point, both analytically and constructively, despite it being characterized by vast differences in privilege, hierarchical structures, epistemological conservatism, and compartmentalization. Such differences must be interrogated and challenged, as several workshop teams chose to do at the UCLA event referenced in the introduction, for example through focusing on "The university as inhumane infrastructure" (figure 1.3).

It can hardly be argued that universities—generally speaking or at all—constitute good examples of humane infrastructures. However, universities are also—at best—wonderous, dialogic, transformative, and experimental

Figure 1.3
Team "The university as inhumane infrastructure" at the UCLA event March 9–10, 2020. From left: Natalie Jeremijenko, Shamell Bell, and Francesca Albrezzi. Photo credit: John Glover.

in unexpected ways. They need to become better, more humane, more caring as well as sharper and braver, but this does not necessarily disqualify them. There are certainly other possible places for capacity building, but what helped center universities in our process was critically discussing some key existing and potential qualities of universities (chapter 4). These included being able to have and build trust, having convening power, and encompassing a range of people, resources, and traditions that could help facilitate transversal knowledge production.[55] Ultimately, what is required is also a genuine wish to be outward facing and to move outside the structure of the university, to be elsewhere with others.

Although the somewhat idiosyncratic journey sketched above is one of extension—from research to civic infrastructure, from things to frameworks, from knowledge production to societal engagement, and from infrastructure to capacity—what becomes particularly clear is how deeply things are entangled. The rest of this book explores this entanglement in different ways, toward humanistic and humane capacity building.

Defining Infrastructure

Infrastructure is an intriguing concept, in part because it does not lend itself to being defined. There is something worrying, almost uncanny, about that which cannot be pinned down linguistically and conceptually—a quality that has followed "infrastructure" for a long time. The term was seen as problematic after the Western Union Defence Organisation and NATO started to use it more extensively in the early 1950s, and while the word is obviously much more established now, its presumed vagueness is often highlighted in contemporary policy discussions in the US and elsewhere.

In a very different context, a scholarly conference on infrastructure in 2018, the participants were asked what three words came to their minds when thinking about infrastructure:

> **The answers ranged from visible, invisible, pickled, informal, relational, and poetic, to material, embodied, pirate, powerful, and circular. A colourful mess! While none of us chose the same term, everybody would surely agree on the answer suggested by Busarin Lertchavalitsakul (Naresuan University), namely "confusing."[56]**

There is value attached to "infrastructure" in this account. It is difficult to define, confusing, and perhaps also wonderous. In any case, it seems likely that the reported discussion was lively and not predictable. Another variety of value attribution can be seen in early newspaper reporting after NATO adopted the term. It is not surprising that the newness of the term was indicated in different ways (including inverted commas), but the word was also made into an object of ridicule. References were made to "intellectual highbrows" who would use such a Latinate term (comment by Winston Churchill), the "heavy handicap" of the term (Dean Acheson, President Truman's Secretary of State), trouble caused by a "warrior-invented word" (*Dagens Nyheter*), and "a bizarre word" (*Frankfurter Allgemeine Zeitung*).[57]

Historian Dirk Van Laak reports on how the *Frankfurter Allgemeine Zeitung* saw the use of *Infrastruktur* as an example of the thoughtlessness associated with "promiscuous use of incomprehensible words."[58] On a somewhat similar note but more recently, "green infrastructure" has been called a "corruptible concept" that can be hijacked by interests.[59]

Categorizing Infrastructure

One reason why infrastructure is a difficult word is likely that its reference is abstract (as well as concrete in context). "An infrastructure" is not likely to bring up a specific reference (even if we may think of let's say a bridge). It is an abstract super category whose limited concreteness can also be seen in the way it is classified linguistically. For example, it is not fully countable (like chairs or pillars). Countability is a strong quality in concrete nouns ("seven chairs," "ten pillars"). "Infrastructure" is more like abstract nouns such as "idea" (which we can count but rarely do) or super category words such as "furniture" (which bring together things that we do not necessarily count together because they are dissimilar).[60] Although it is possible, we rarely talk about "three infrastructures," especially not counting things that are different, such as a bridge, a tunnel, and a road.[61] This abstractness combined with the tangible physicality of much infrastructure contributes to making it into a complex concept.

Another reason why the term is challenging is that infrastructure is not just "out there" (as noted) but rather can be regarded as being epistemologically, politically, and historically inflected; think of the learning ideals performed by online learning platforms or the bias of many health-care systems.[62] Infrastructures also tend to have certain qualities, such as being thought of as foundational, instrumental, standardized, public, systemic, and material, but not everything with those qualities will be classified as infrastructures. And not everything considered to be an infrastructure will have most of those qualities. The "same" artifact or system may be classified as an infrastructure now but not in the past or in one context and not in another, the reason being that it is not really the same thing *infrastructurally* in these different contexts and situations. This makes it practically impossible to answer the question "What is infrastructure?" I follow Brian Larkin in seeing acts of defining infrastructures as categorizing moments.[63] Such categorizing moments, I argue, are not just relevant analytically but can also be thought of as a constructive modality, an opportunity for change.

Some things are more likely than others to be called infrastructures consistently, such as transportation systems (roads, railways, bridges, etc.), urban structures (e.g., buildings and sewage systems), and electricity (power networks, power supply, outlets). It is almost impossible to imagine a conception of infrastructure that would not include such phenomena, and these prototypical instances will be frequently listed in definitions. As information studies scholar Christine Borgman notes in her study of cyberinfrastructure policy, infrastructure is often defined through such examples.[64] When research infrastructure gets defined, for instance, there is almost always a list of examples indicating what infrastructure can be (including labs and computational resources) and—by exclusion—what infrastructure cannot be or would not normally be. Sometimes, even a single example will do. Common in infrastructure narratives, the inclusion of such project lists helps categorize and define infrastructure in that specific context.

Infrastructure is often taken for granted despite cursory or even missing definitions. It is perhaps counterintuitive that the lack of conclusive definitions of "infrastructure," related to the concept's complexity and abstract materiality as discussed above, can help explain the vitality and extensibility of the term. As suggested by cultural anthropologist Ashley Carse, infrastructure can be seen as a keyword, like "culture" and "democracy." Such socioculturally embedded words are often contested and difficult to pinpoint.[65] Cultural theorist Raymond Williams saw keywords as important binding words that reflected a changing world and connected to "certain forms of thought."[66] Along similar lines, as noted earlier, infrastructure can also be seen as a boundary object, bringing different actors and interests together. Boundary objects are adaptable to different perspectives or viewpoints but have enough integrity to maintain their identity across them.[67] I argue that this rich, indeterminate, and boundary-crossing quality gives "infrastructure" a potentiality that makes it useful for considering ways of changing structures through collectively pushing them gently as well as more radically.

Useful Adaptability and Gradual Extension

A historical example of how this plasticity can be advantageous is provided by the aforementioned adoption of infrastructure by NATO, particularly through the Infrastructure Committee, which was established in 1951 and is now the longest-lasting NATO committee.[68] Infrastructure was defined as "the static items of capital expenditure which are required to provide the material support for operational plans necessary to enable the higher command to function and the various forces to operate with efficiency."[69] Some examples were given, including headquarters, airfields, and pipelines. This was not a precise definition, and in fact, the committee is retrospectively clear about the usefulness of such definitional flexibility:

> **The lack of a rigid definition is actually one of the strengths of the programme. There is a flexibility and diversity to the programme which has served the Alliance very well.[70]**

As historian Sanne Aagaard Jensen has shown, however, there were negotiations and efforts over time within NATO to narrow the definition down and to make it more specific.[71] A working group appointed to produce a stricter definition came to the conclusion that a more rigid definition would not be helpful and would risk having to include installations that were not considered suitable for common infrastructure financing, likely resulting in too many infrastructure projects. It would also seem that the opposite might have been true. The flexibility of a relatively loose definition of infrastructure made it possible not only to exclude things that might otherwise have been prescribed but also to include things that may not have qualified as infrastructure, given a more precise definition. Negotiation was an important part of the process, and again, this example points to how difficult it often is to establish literal workable definitions of infrastructure. Changing a definition or framework can be important, but the act of categorizing infrastructure often occurs in complex processes between multiple actors.

Aagaard Jensen, whose study centers telecom technology in Denmark, argues that the NATO Infrastructure Committee work helped demilitarize the term "infrastructure," as it became difficult to separate military and non-military infrastructure as well as national and international infrastructure.[72] The introduction and spread of the term over time are obviously relevant factors when considering how it gets defined and categorized. There was a limit to what could reasonably have been categorized as infrastructure in the early 1950s when the military reading was being established and the term as such was relatively rare. It is noteworthy that the first US president to use the term—as recorded in public documentation in the American Presidency Project—was John. F. Kennedy in 1962, referring to the military base Quonset Point being part of "a great infrastructure along the Atlantic coast."[73] It is one of four instances in the full John F. Kennedy corpus. Two occurrences directly relate to military infrastructure, one to engineering and educational infrastructure needed for defense contracts, and one to gambling infrastructure (which had been shut down) when talking to US attorneys about the "war against crime."[74] We can see semantic extension at play here as the use and range of "infrastructure" extends gradually over time.

Quantitative data are revealing. The presidential use of "infrastructure" rises after 1980 (Ronald Reagan). A surge started in the early 1990s with Bill Clinton and then continued with George W. Bush (with four and six times the numbers for Reagan, respectively). Barack Obama's numbers were more than one and a half times as high as those of his two immediate predecessors together.[75] Infrastructure was not the same thing during all these presidencies, and it is now unthinkable that US presidents would not shape their agenda and discourse around infrastructure and, in doing this, also be part of shaping infrastructure. Consider, for example, Donald Trump's nationalistic and privatization-driven vision of infrastructure in the 2016 election as part of his America First political agenda.[76]

We need to be aware that the range of the word "infrastructure" varies over time and that adaptation and extension are gradual processes. It took time for infrastructural forms of thought to get established, and they change over time, which is also likely to apply to new developments. Imagine looking in a dictionary and not even finding the word as reported by this commentator in the late 1980s:

> **The emergence of "infrastructure" as a generic concept and prominent item on the public agenda is a phenomenon of the eighties. The word itself does not appear in leading dictionaries of the seventies. (Those I have checked are *The Oxford English Dictionary*, *The American Heritage Dictionary*, and *The New Webster Encyclopedic Dictionary*.)**[77]

Dictionaries are not indisputable as evidence for when a word came into a language and how it was used, as they come in many versions, sizes, and timelines and rely on different types of materials. They also reflect a time lag, and examples, especially in earlier times, can be anecdotal or inherited from other dictionaries and are often presented without much context.[78] In any case, there were dictionaries that included the term earlier, both as a specialist term and in the more general sense.[79] What the above quote illustrates is a sense of ongoing change, both as perceived by an individual writing a review of two volumes on infrastructure investments and as codified in a few dictionaries (perhaps found in his own office). Tellingly, one of dictionaries referenced—*The American Heritage Dictionary of the English Language*—featured a comprehensive definition of the term in the 1980 edition, whereas it was not included a decade earlier (in the 1970 edition). Another dictionary, *The Random House Dictionary of the English Language* from 1966, picked up the term earlier but with a military definition: "military installations and facilities belonging to NATO." These changes were about the word not only becoming more accepted but also increasing its territory, being used in more contexts.

When considering linguistic representations of change, it can also be helpful to consider quantitative data, including frequency information. Looking at Swedish materials, one source aggregated consisted of two daily newspapers (*Dagens Nyheter* and *Svenska Dagbladet*) and the two leading tabloids (*Aftonbladet* and *Expressen*). In this corpus, there are a total of 118 newspaper pages containing the word *infrastruktur* between 1950 and 1970.[80] These are absolute numbers, which should be treated cautiously, as they are not adjusted for the amount of text. Also, a newspaper corpus such as this one does not represent all written and spoken language. Nevertheless, it seems clear that the word *infrastruktur* was not really part of the general vocabulary from 1950 to 1970.[81] Looking at the following decades in the same newspaper corpus, there was an increase in use in the mid-1970s, but numbers were still low (a total of 504 instances for the 1970s). Then, usage more than doubled in the 1980s (1,276 instances) and increased by a factor of six in the 1990s (7,692 instances) followed by a more marginal increase in the 2000s (8,620 instances).[82]

Parliamentary motions provide another perspective on the expanded use of the word in the Swedish context. In an archival dataset going back to 1867, there are 194 documents in total containing the word until 1984 followed by a sevenfold relative increase until 1991 on an annual basis, where the usage of the term leveled out before a slight increase in the early 2020s.[83] The first recorded instance of *infrastruktur* in this material (motions) is from 1961 and relates to support for developing countries (used in the definite form and preceded by "what is usually called" to indicate the term's newness).[84] Another indication of the adaptation and significance of the term is whether it was included in the titles of such motions. Looking at data from two five-year periods, 1971–1976 (forty-seven occurrences of the term) and 1981–1986 (212 occurrences), there are no titles containing the term.[85] However, in the material from 1991 to 1992 (513 occurrences), there are nineteen titles containing *infrastruktur*. These data point not only to the increasing centrality of the concept but also to a relatively slow establishment process.

Shifting Frames

These data demonstrate that "infrastructure" was established more broadly in the decades around 1990 and also that a broad core meaning of the term was solidified in this period. An important factor was a large-scale shift in the Western world from public works monopolies to deregulated infrastructural systems and a global capital infrastructure market driven mainly by perceived insufficiencies with the regulatory system and the rapid expansion of an ideological free market agenda. The rail industry was an early case that had bipartisan support in the US, and several acts were passed between the mid-1970s and early 1980s.[86] There are two interesting points to be made here: one to do with how "infrastructure" came to supplant and gradually replace "public works" and another to do with how a strong framing of infrastructure as failing emerged.

Words and concepts do not exist by themselves but rather relate to other words and concepts. Choosing between "public works" and "infrastructure" can be a meaningful choice, as we saw in Laura Bear's appeal for a return to "public works" mentioned in the previous section.[87] According to semantic or lexical field theory, the lexicon is "organized along conceptual affinities and contrasts reflected in the semantic relations that structure fields."[88] Words can be said to be part of semantic or lexical fields with related words (similar, partly overlapping but not "the same"), and while this may not be a precise approach, it can provide a useful perspective. Looking at "infrastructure" and "public works," frequency data from the Corpus of Historical American English (COHA)[89] suggest that the two terms had opposite trajectories over the period 1950–2000, with "infrastructure" on the rise and "public works" on the decline (figure 1.4).[90] The two paths intersect between 1970 and 1980 toward the end of the decade.

While we should be careful to draw far-reaching conclusions from this type of material, the trends point to how the term "infrastructure" gained

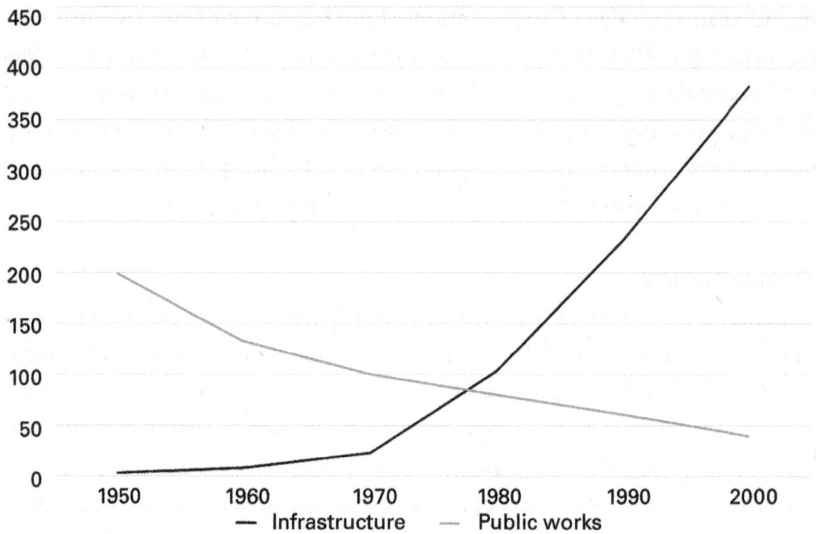

Figure 1.4
Trajectories based on data from the Corpus of Historical American English (instances).

prominence in the latter part of the twentieth century, especially from the late 1970s onward.[91] No instances exist in the corpus for usage of "infrastructure" prior to 1950. The related term, "public works," goes back to the 1820s in the material, was most frequent between 1880 and 1970, and shows a substantial decline between 1950 and 2000. This pattern would seem to point to a gradual turn away from "public works" as an idea and frame to alternative frames, including "infrastructure" around the late 1970s, related to a set of societal and cultural shifts, including deregulation.

We can see this shift taking place in the early 1980s exemplified through the book *America in Ruins: The Decaying Infrastructure* by Pat Choate and Susan Walter.[92] The original subtitle of Choate and Walter's work (when published as a report) did not include "infrastructure," but the word was added to the book title that was published two years later (thus replacing "Beyond the Public Works Pork Barrel").[93] The book itself mostly references "public works" rather than "infrastructure." In the first thirty pages of the book, there are 103 instances of "public works" and a total of six instances of "infrastructure/s."[94] Another common term in the sample is "public facilities" (nineteen instances). It can be argued that the variation in this material points to an ongoing shifting of frames (as well as a publication that underwent several iterations).

On the cover of *America in Ruins*, infrastructure is epitomized by once-strong stone buildings crumbled and abandoned in the detail from Giovanni Battista Piranesi's *View of the Campo Vaccino* from 1772 (figure 1.5). The cover also demonstrates another infrastructural attachment or frame, which perhaps would not be found in a dictionary definition but has become part of the idea of civic infrastructure: crumbling infrastructure.

Choate and Walter observed the extensive neglect of public works projects in the US (roads, water, etc.) and contributed to a framing of civic infrastructure as falling apart (aging, decaying, and crumbling) and in neglect (infrastructure famine and deficit). Other observers made similar points around this time and in the following decades, and there can be no doubt that there has been a decline in the quality of infrastructure and a partial shift toward looking back rather than forward (illustrated by the etching below, which records the glory of ancient Rome and alludes to the past glory of America).[95] Choate and Walter's work has been critiqued for being opportunistic, alarmist, and federalist,[96] but crumbling infrastructure continues to be an active and frequent framing (and reality) for governing and policy work—making infrastructure—as is evident from contemporary social and societal discourses and policy (and not only in the US). For example, the grading of risks that Choate and Walter carried out for city infrastructures in a scale of twelve grades (from "poor and deteriorating" to "very good" in that order) is reflected in the report card that the American Society of Civil Engineers (ASCE) issues every four years.[97] The 2017 overall grade was a D+, as manifested in the society's information materials (figure 1.6).[98]

The score reflects ASCE's professional assessment of the situation, which is undoubtedly dire. The combined effect of the grade and the visuals is to categorize and define "infrastructure" with a focus on the imminent failure of physical infrastructures. Environmental historian Thomas Zeller describes the report cards as "an instrument of professional politics," since engineers have most to gain from further investments in this area.[99] The video clip highlighted on the website is a fly-through, with virtually no people (apart from a few male construction workers), starting out from a city and containerships and moving through forest on a train to the bridge in green water.

Figure 1.5
Cover image of *America in Ruins* (1983, Duke University Press). Image credit: Duke University Press.

Figure 1.6
From the American Society of Civil Engineers report card video for 2017. Image credit: ASCE.

Most of the materials presented in this and the previous section have emphasized the expansion and leverage of infrastructure as a term and concept. It does not seem likely that "infrastructure" will lose its status any time soon, given its institutional entrenchment as well as its relative adaptability and potential moldability. However, it may lose traction over time—for example, through becoming less useful as a policy frame. Also, ideas and frames can shift over time, as we saw with "public works." Other frames may become stronger, some perhaps building on different conceptions of the world. For example, it is not inconceivable that "research infrastructure" could lose its central place in much of Europe if universities moved away from instruments, facilities, and capital investments as an organizing principle for higher education, perhaps instead centering students and education, excellence (regardless of infrastructural platforms), large research programs with blanket funding (where instrumentation would be up to the research team), or far-reaching resilience.

Research Infrastructure

In the Swedish Research Council (Vetenskapsrådet) annual report from 2001, there is not a single instance of the word *forskningsinfrastruktur* (research infrastructure), although there are five instances of *infrastruktur* (infrastructure) in a more general sense (eighty-four pages in total).[100] In the equivalent report from 2022 (153 pages), there are 86 instances of *forskningsinfrastruktur* and eighteen instances of *infrastruktur*.[101] This very simple measurement illustrates the influx of infrastructure thinking into the realm of the principal state-run funding agency in Sweden and higher education more generally. What happened over these years, and what does it mean?

It may also be useful to consider what did not change. For example, it is not the case that there was no funding in 2001 from the Swedish Research Council for equipment and facilities of the type that would now be included in research infrastructure as a category. The 2001 budget had funding allocated for so-called expensive equipment, longitudinal databases, and few international resources (including a contribution to the European Organization for Nuclear Research [CERN]). However, while this commitment represents a substantial level of funding, it is nowhere near the massive investments made twenty years later. For 2022, the cost for research infrastructure is about the same as for regular project funding for all disciplines, which is the staple of national research funding.[102] Taken together, research infrastructure and regular project funding represent about three quarters of the total research budget for the Swedish Research Council for 2022.

This development is complex, reflecting a set of changes including extensive digitization of research processes and data management (increased infrastructuralization), but one indisputable factor is that "research infrastructure" as a concept and frame has become an organizing principle for higher education. A group of Swedish university presidents argued in an opinion piece in 2017 that research infrastructure and its funding is a key issue for the future of Swedish universities.[103] Research infrastructure is integral to Horizon Europe, the European Union's ambitious funding program for research and innovation for 2021–2027, and in a recent budget request, the US National Science Foundation describes research infrastructure as "foundational to the scientific endeavor" and an important enabler for the overall goals of the NSF.[104]

The Establishment of Research Infrastructure

Around the mid-1990s to early 2000s, "infrastructure" became a way of framing academic research tools, facilities, and structures in Canada, the US, Europe, and elsewhere, often as part of a national or pan-national agenda.

Research infrastructure, as articulated in many policy documents, is not only about scientific capability and excellence but also about competitiveness at different levels (e.g., Sweden's or Europe's competitiveness), innovation potential, and managing and sharing costs. There is often a sense of transformative change in a proximate future (not quite there but within reach) and a need for urgency in policy documents.[105] For example, in an important NSF report from 2003, it is stated that the cost of not investing in cyberinfrastructure is high, in terms of both lost opportunities and "balkanization of the research communities,"[106] and another early report predicts that in five to six years, if the recommendations are implemented, humanities and social science researchers "will be answering questions that today they might not even consider asking."[107] Such reports are full of what may be called "infrastructure speak," which reflects the properties and ideology of infrastructure work. For example, the innovative and futuristic associations of infrastructure are manifested through language that almost inevitably points forward and promises radical change for which infrastructure plays a central role. It has been noted that there are 133 instances of "new" in the eighty-four-page NSF report on cyberinfrastructure from 2003 mentioned above.[108]

Gradually, research infrastructure has become an important controlling mechanism, policy factor, and strategy for higher education. It brings out many of the issues discussed in this chapter, including the effectiveness of open-ended definitions, the strong technological imperative, and the inflection of infrastructure. This section prefigures the discussion of humanistic engagement with infrastructure in chapter 2, including how the humanities have negotiated research infrastructure. It also considers the framework and mechanics established, which will be useful for the discussion on challenging and shifting infrastructural frameworks in chapters 3 and 4.

Research infrastructure thus provides a relatively contained example of how the introduction of "infrastructure"—as part of a range of broad societal and cultural shifts—can be instrumental in influencing and changing a societal sector—a sector that, in itself, is central to this book. Again, it is not that most of the things (laboratories, microscopes, facilities, etc.) that are now called "research infrastructure" did not exist earlier but rather that "infrastructure" emphasized a form of thought, a method for putting things together, and a way to think about and organize higher education—not entirely new or all-encompassing but far-reaching.

The Canada Foundation for Innovation (CFI), which was instituted in 1997 and preceded by a longer sector-wide process, represents one of the earliest implementations of "research infrastructure" as a major funding and policy category.[109] This is how research infrastructure is currently defined by the CFI:

> **The infrastructure funded by the CFI includes the state-of-the-art facilities, laboratories, equipment, computer hardware, software and databases necessary to conduct leading-edge research.[110]**

Definitions of research infrastructure tend to specify high-quality facilities, tools, "data," and technology required to support or generate world-class research in this way. As a category, research infrastructure opens up a politicized and institutional space for negotiation. Just like NATO infrastructure, it is difficult to pin down, which can be seen as a problem.[111] Definitions are rarely conclusive or, for that matter, meant to be conclusive, but they are also not meaningless, just political, or hype. Not everything will qualify as infrastructure or qualify as easily. For example, research infrastructures tend to be technological, instrumental, generic, long term, and large scale. They are also epistemological, at least to some degree. The policy level, however, is often supercharged. There are assumptions built into definitions and the associated policy discourse. The CFI adds that research infrastructures are what "researchers across all disciplines need to think big and innovate," and they also help researchers "push the boundaries of their work."[112] Infrastructure is given clear agency and transformational capacity, which is a typical feature of research infrastructure policy. A more specific example can be offered from a description of the European Spallation Source, which is likened to a giant microscope and given considerable agency through the visualization capabilities of the tool:

> **Improved visualization techniques enable researchers to observe our world and universe better. From the very large to the very small, when science moves forward, it is often due to breakthrough improvements in the tools available for scientific experiments.[113]**

These tools are taken to have immense capability, and improved instrumentation is what leads to science moving forward. As noted earlier, detail and precision are often added through examples in policy documents. A list given by the CFI includes "ultra-modern research spaces" (e.g., advanced imaging facilities, genome centers, storage hubs), DNA sequencers, mass spectrometers, and high-precision lasers.[114] A visual (the thumbnail of a video) shows a science lab environment.

Looking at the CFI, which must be regarded as a representative example of research infrastructure initiatives, we can observe several fields of tension, including between the transformative aspirations and the actual instruments and their "down-to-earth" role in research, between the abstract category of research infrastructure and the physical equipment, and between a commitment to a broad range of disciplines and the arguably non-inclusive nature of the definition and examples listed (few humanists would identify with genome centers and high-precision lasers). Furthermore, the emphasis on technology and instruments provides an epistemological frame that is more compatible with some epistemic traditions, especially tool- and data-centric modalities, than others. There is a risk that such frameworks can hinder alternative out-of-the-box conceptions of research infrastructure, not only epistemologically but also institutionally and politically.

Infrastructure policy work is about shaping. The basic question of what counts as research infrastructure is not a given but reflects rules, priorities, and epistemic predispositions that have developed under specific conditions and that are exacted in acts of categorization such as funding calls and decisions. These have become encoded in infrastructure policy and funding agency priorities. For example, policy experts Débora Lopreite and Joan Murphy note how the Canada Foundation for Innovation has established a set of rules associated with research infrastructure, which include requirements on universities to submit strategic financial plans and secure investments from other funders.[115] The CFI also designates infrastructure as capital cost rather than operation costs as part of an innovation and technology framework.[116] This strategy, in turn, puts pressure on other institutions to supply operational funding, in fact shifting the overall logic toward infrastructure, innovation, and top-down strategy.[117] On the other hand, it is important to acknowledge that the CFI, as part of a complex ecology of institutions and actors, has been able to absorb costs that might otherwise have been difficult to cover, and the long-term perspective of research infrastructure also has benefits compared to short-term university or research project allocations.[118]

One of the key properties of research infrastructures is large scale and longevity. They are big, long term, and costly. The budget for the European

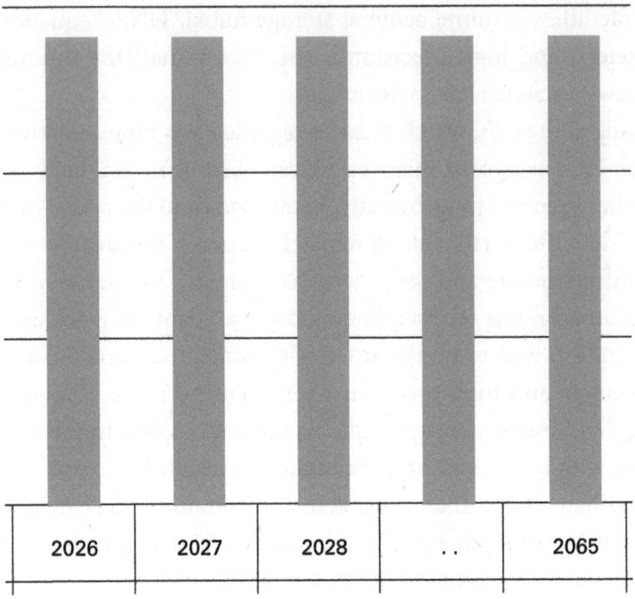

Figure 1.7
From the ESS budget (taken from the EU Statutes, 2015). Image credit: ESS.

Spallation Source, a very large European scientific infrastructure, extends from 2013 until 2065.[119] The semi-permanent nature of this infrastructure is indicated by how the term "steady state operations" is used to describe the long-term running costs (from 2026, at one stage, this phase was costed at EUR 5,600M).[120] The remarkable scale is perhaps most strongly manifested by the elliptical ". ." in the budget graph (figure 1.7), encapsulating thirty-six years of operation.[121]

There had been unified European-level support for providing access to large-scale facilities such as large installations and instruments earlier (beyond state-based collaboration regarding individual installations such as CERN), but it was not until the so-called Fifth Framework Programme (1998–2002) that "research infrastructure" was specified under that name and given a more central role in the European context. Still, the focus was on access and improvements needed for access, since research infrastructures themselves were seen as a national responsibility.[122] In January 2000, "Toward a European Research Area" was published by the European Commissioner responsible for scientific research.[123] In this document, which gained broad support, research infrastructure is given a central role in the vision of a European Research Area (ERA), a kind of common market for research and development, and developing a more far-reaching European approach to research infrastructure, including creating new installations, is advocated. One part of the continued process was a conference that took place in Strasbourg in September 2000.[124] The European Strategy Forum on Research Infrastructures (ESFRI) was established shortly afterward in 2002. ESFRI and the creation of an extensive European framework for research infrastructure were influential for national developments across Europe, including the establishment of the Committee for Research Infrastructure at the Swedish Research Council (KFI) in 2005.

In the first European road map to infrastructure from 2006, European research infrastructure is largely framed in relation to pan-European knowledge growth and capacity building and creating a "research and innovation area equivalent to the 'common market' for goods and services."[125] European research infrastructure is thus bound up with the European project and its goals, including the single market, and becomes a way of manifesting Europe, European identity and integration, and European strength. Cooperation is a key sentiment for European research infrastructure, also in a longer "pre-infrastructure" trajectory. Big science is costly, complex, and challenging and provides a strong case for the power of European cooperation and for an integrated Europe as idea.[126] Facilities such as CERN and the European Space Agency (ESA) provided convincing examples. Cultural anthropologist Stacia Zabusky shows how national identities, professional roles, and backgrounds come together in ESA as a structured unity around missions, instruments, and technology:

> **ESA is not simply a technical agency, then, dedicated to science and high technology to the exclusion of all else; it is an active participant in the political and cultural arena where European unity is being constructed.[127]**

There was a robust development of European "research infrastructure" following the establishment of ESFRI in 2002, as evident in the introduction of national road maps across the member states.[128] It is not surprising that national research infrastructure policy was different than European-level efforts, given the newness of the category, different priorities, and the fact that the member states had to provide the funding. Over time, there has been gradual homogenization, however, as a European framework developed. Nevertheless, European research infrastructure logic continues to be complex and contingent on buy-in from member states under the soft intergovernmental "open method of coordination" according to which member states can agree on (and fund) policy in certain areas.[129]

Road-Map Mechanics

The particulars of policy implementations play a formative role not only for the development of research infrastructure but also for the academy more generally as infrastructure becomes a modus operandi. This is particularly true of the European case. Infrastructure work comes with mechanics that operate across different levels and that can come across as self-replicating. I will now look at one example of such mechanics—the road map—in some more detail. It is a key framing device or tool for infrastructure work that outlines current and future priorities and priority potentials.[130] The origin of the road map can be traced to industry use, including the car industry, and to what is called the "technology road map."[131] Multiorganizational road maps are said to promote knowledge sharing and facilitate "the development of a collective vision that can lead to action and collaboration."[132]

The importance of road mapping can be related to the presumed scale (cost, longevity) of infrastructure, and road maps attach to a financial, administrative, and innovation-driven logic. The argument is that road maps are needed for investment decisions, since there is limited funding and competition between different initiatives and because infrastructure investments are long term.[133] Additionally, the complexity of something such as research infrastructure creates a need for systematicity. This logic supports a particular model of research infrastructure as discussed above, and there is a strong normalizing path-dependency effect at play.

The tendency to normalize is evident in the first road map on the European level, where it is stated that some fields of science—not as well structured and organized as the others—should be "stimulated to integrate further and be more proactive at European level so that they can become more effective in the future."[134] In an OECD report that discusses road-mapping work, the authors claim that it cannot be assumed that "communities are naturally inclined towards this type of introspection without an external stimulus that a road-mapping exercise provides."[135] It is easy to see how road mapping becomes a mechanism for control and administration with its own separate logic.

Such logic has to be maintained: road maps have to be managed in order to function. There have to be processes of prioritization, resourcing, and harmonization. The Swedish Research Council has a procedure in place for taking in new proposals, ranking them, inviting proposals, prioritizing funding, and producing a road map (or guide). There also have to be contingencies for disengaging from infrastructures and removing them from road maps or, as a representative of the Swedish Research Council put it in a seminar, ceasing funding.[136] Road maps at different levels have to be coordinated, and there is consistently a push to harmonize investment. In an ESFRI report on the long-term sustainability of research infrastructure, national bodies are asked ("invited") to synchronize and harmonize their research infrastructure investment strategy with a "pan-European vision by taking advantage of the landscape analysis and road mapping procedures developed by ESFRI and other players at the global level."[137]

Emergent Infrastructure Logic in Sweden

As noted, road mapping imposes an iterative logic, which is prominent in European infrastructure work. For example, national and European-level infrastructure policies filter down to individual universities in Sweden.[138] The establishment of the Committee for Research Infrastructures in 2005, which became the Council for Research Infrastructures in 2008, has played an important role in creating a national incentive structure for infrastructure. Following the establishment of a national funding scheme and the workflows implemented—including four infrastructure guides (2006, 2007/2008, 2012, and 2014) and a substantial appendix issued in 2016 as well as new infrastructure guides/editions in 2018 and 2023—most institutions of higher education in Sweden have undergone processes of inventorying research infrastructures and research infrastructure needs starting around 2010. This inventorying work is also related to a changed process for handling research infrastructure nationally, essentially making local infrastructure the responsibility of their institutions.

At the local level, universities in Sweden have been instituting processes and frameworks building on national and international frameworks but with considerable variety and different levels of transparency. Looking at the situation at the end of 2019, almost fifteen years after KFI was established, most university websites had a category for "research infrastructure" under "research" but as a fairly recent addition. Among the six big comprehensive universities in Sweden, Stockholm University introduced the category at the end of 2014, Gothenburg University at the beginning of 2016, Uppsala University at the beginning of 2018, and Umeå University at the end of 2018. Lund University and Linköping University still did not have a clear infrastructure category under "research" (or elsewhere). This was still the case for Linköping University in December 2023, although for Lund University, the large infrastructures ESS and MAX IV had been given their own menu heading under "research" (introduced in February 2023).[139] The fact that research infrastructure is now often listed alongside prioritized research areas, research projects, research groups, and graduate education on the university websites demonstrates how the landscape of research in Sweden has changed over the last decade. The organization, discourse, and understanding of research in Swedish higher education have become dependent on research infrastructure as a category.

Most of the websites explored (as of December 17–18, 2019) present research infrastructures in the form of project or infrastructure lists. Because research infrastructure is often defined by examples, the specific composition of such lists is quite significant. In contrast to national or international definitions, lists at the level of individual universities are closer to the ground, as they are likely to reflect university policy directly. At Uppsala University, essentially all the infrastructures listed were in science, technology, engineering, and mathematics (STEM) disciplines. The only exception out of sixteen infrastructures in the list—filed under a third-tier category of other strategically relevant infrastructures—was the university library.[140] At Stockholm University, both human sciences and science infrastructures are listed, and it is pointed out that there are no sharp definitions of what counts as infrastructure.[141] Consequently, the Stockholm lists seem very inclusive, whereas Uppsala University is more restrictive. Umeå University has a more controlled selection and a database structure.[142]

In general, the websites show compliance with national and European infrastructure policy and thinking but with local variations also manifesting different stages of entrenchment and institution-specific negotiation of research infrastructure. Whether research infrastructure is given attention on the websites or not, most universities will now have procedures, policy documents, committees, and budget allocations for research infrastructure. Additionally, many institutions of higher education in Sweden—including all the aforementioned institutions at the time of investigation—have a deputy vice chancellor, a pro-vice-chancellor, dean, or the equivalent with a responsibility for infrastructure questions.

It has been noted that infrastructure investments need to be balanced with investments in research,[143] but also that infrastructure thinking in higher education moves decisions away from the universities. This is one reason why Swedish universities operate through a reference group for research infrastructure (URFI), which currently consists of representatives from eleven universities and the national organization for universities (SUHF).[144] URFI does not represent all universities, however, and a large number of newer universities and university colleges are excluded.[145] URFI noted in 2020 that 84 percent of expressions of interests for infrastructures to the Swedish Research Council in 2019 came from their ten universities, some from other institutions, and only 5 percent from other universities,[146] which not only becomes a way of rationalizing URFI as an organization but also reinforces the big-size template of infrastructure. Only established "big" universities belong to the landscape of research infrastructure whose stability is guaranteed by what very much comes across as a self-generating process.

Such patterns are conducive to infrastructure work, where large size and established structures are typically (and definitionally) at the forefront. It is unsurprisingly easier for already well-funded and well-supported institutions to be and become infrastructural. There are layers of privilege and templating to overcome for a small institution with limited resources, such as a university college or small university in Sweden (still obviously privileged) or a community college or underfunded state university in the US, even more so if the infrastructural conception does not conform to intra-institutional or national templates. Similarly, different provincial regions in Canada have different ability to leverage resources for infrastructure proposals.[147] ESFRI's road map is heavily skewed toward Western and Southern Europe, especially if the distribution of leadership of projects and landmarks is considered as well as the total number of engagements of each country.[148]

Overall research infrastructure as a framework is predominantly a Western-world phenomenon, and just as with research (and other) resources more generally, there is vast inequity.[149] It is not enough to give researchers from low- and middle-income countries access to research infrastructures in infrastructure-rich high-income countries, but there must be room for

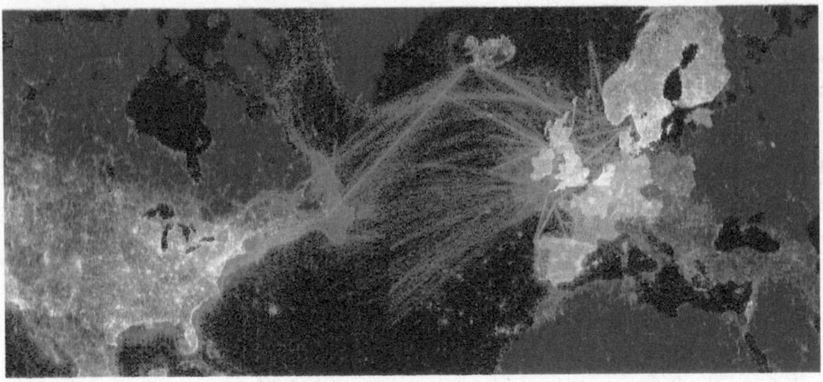

Figure 1.8
From the cover of the Swedish Research Council Guide to Research Infrastructures 2014. Credit: The Global Biodiversity Information Facility.

substantial local development.[150] Also, with an established elaborate framework such as research infrastructure, not least in the European context, there is a risk that not only research infrastructures but also the ideas and logic associated with infrastructure get exported to other parts of the world. The competitive inward-focused tendency of research infrastructure does not help. It is telling, for example, that one of the directions in the new research infrastructure work program for Horizon 2023–2024 (EU funding), which includes international aspirations, has the name "Developing, consolidating and optimising the European research infrastructures landscape, *maintaining global leadership*" (my italics).[151]

Even if we should not read too much into individual images in presentation materials, it is somehow telling that the cover of the Swedish Research Council Guide to Research Infrastructures from 2014, a species occurrence visualization taken from the Global Biodiversity Information Facility (figure 1.8), mainly covers Europe and the US.[152]

Not only is the cover image clearly about scientific data (as are most of the Swedish guides to research infrastructure 2006–2016) and a specific mode of visualization, but it is also a visualization that has been cropped to focus on one part of the world.[153]

One parameter that makes infrastructure work especially challenging for smaller or inexperienced institutions is that there is a considerable administrative and processual overhead from the very beginning, especially in systems such as the European one. It takes time, resources, and knowledge to get a sense of the landscape, form consortia, secure local support and traction; to get national support; to place and negotiate a suggested infrastructure in the broader international landscape; to write proposals that are infrastructural enough; and to go through the continued process if selected (and this is no conclusive list). Although there is a great deal of top-down and political work in such processes, it is important to emphasize that most research infrastructures are not just political and visionary but also epistemological and the result of complex processes that involve researchers as well as administrators, government representatives, and policymakers.

Nevertheless, administering a research infrastructure can be a significant burden, especially given the various conditions and measurements introduced at the European level. There is a pressure, for example, to synergize efforts across initiatives. Humanities-based EU infrastructures such as CLARIN ("the research infrastructure for language as social and cultural data") and DARIAH-EU ("the pan-European infrastructure for arts and humanities scholars") are infrastructural anchors, but even so, they are amenable to consolidation efforts.[154] In a European Commission report, DARIAH is asked to "fully exploit synergies with other Research Infrastructures in the SSH domain (for example CLARIN) and periodically assess the opportunities for closer collaboration and/or integration between them."[155] This dependency on synergies and big size on the national and pan-national levels makes the threshold for new initiatives high.

Another example is an elaborate mechanism—still under development—for keeping track of progress and monitoring existing research infrastructures. An important metric for such monitoring is provided by so-called key performance indicators (KPIs), whose identification across infrastructures can be challenging according to a report.[156] The report gives attention to narratives as well as metrics, which is laudable, but the bottom line is that things need to be measured under this logic. For example, this is the first key recommendation in the report:

> **All KPIs should be aligned with the objectives of RIs and fulfil RACER criteria: Relevant, Accepted, Credible, Easy to monitor, Robust. Each KPI should be accompanied by a reference sheet that provides a definition, data source(s), method of calculation, and other information concerning calculation or applicability.**[157]

Just like with definitions of research infrastructure, this recommendation embeds and imposes a set of assumptions, including that research infrastructure performance must be calculated. More generally, the emphasis on performance is tightly linked to a functional view of infrastructure, which, in the case of research infrastructure, can be translated into metrics such as number of users, number of user requests, and number of publications. There is some maneuverability, however, and it is also up to research infrastructure representatives to extend and tweak the framework by coming up with their own metrics and narratives. Such negotiation, which must include aligning and framing their own work, may contribute to a long-term shift of the monitoring mechanism as needed.

Monitoring and various other mechanisms—including road mapping and definitions—contribute to an infrastructural regime. Research can be said to be part of a new public management logic, which connects to modernist models of governance characterized by standardization, management, instrumentalism, and systemic control.[158] Indeed, research infrastructure is a powerful manifestation of new public management—almost a perfect exemplar—through its role in reforming higher education in a managerial, auditing, innovation/technology-driven, and homogenizing direction. Research infrastructure as an idea and framework brings order, predictability, and incentives to higher education through a set of conditions and mechanisms.

The Beauty of Science

There is a striking tension between this orderly framework, which seems to incorporate a rationalist logic and "pure" notion of science, and the way that research infrastructures are often presented. I will conclude this section by considering an image from the Swedish Research Council Guide to Infrastructures from 2012.[159] It evokes the European Spallation Source in the context of other planned and existing infrastructures in a visualized proximate future. The visual has a prominent position in the guide, filling the third page of the initial three-page executive summary. It was not possible to secure permissions for this image, but ESS generously supplied another image from a somewhat later stage in the process (2014)—darker and less spectacular but with many commonalities (figure 1.9). This type of image often has a futuristic glitz or shine that can be paralleled with the aesthetics of architectural project presentations or techno-scientific visions.[160]

Figure 1.9
ESS image from 2014. Credit: Henning Larsen Architects.

As noted earlier, ESS is a very large Swedish and European investment, and the image used in the Swedish Research Council report reads like part of a political campaign for ESS. At the time of publication of the guide, there was no legally binding decision to build ESS,[161] and the caption (overlaid on top of the lower part of the original image) talks about preparations *underway*, a *planned* installation, and an *expected* date for construction. At this point in time, there were still many issues—some of them prevailing—concerning the use, scientific rationale, national priorities, and costs of the infrastructure.[162] ESS is organized as a consortium, and the Swedish Research Council does not own the facility, although they have an important function in funding it.

The site of ESS is Lund in the South of Sweden, and the bridge to Denmark and the continent can be seen in the background of the image, symbolizing the European perspective. The sun hovers prominently over the horizon (more so in the image in the report) and contributes to the sense of a proximal future and promised scientific and societal progress. Even though the facilities are situated in a scene, there is a sense of purity and displacement, which is supported by the accompanying pastoral landscape. This juxtaposition of nature and technology is in line with a pastoral tradition where technology came to be part of depictions not because of its revolutionary or industrial connection but rather because there was productivity, form, and pride associated with technological manifestations.[163] ESS is an object of pride. There is a spectacle here that can be described as an example of what David Nye calls the technological sublime, "an essentially religious feeling, aroused by the confrontation with impressive objects."[164] The beauty of the scientific is stressed by the ESS written materials, which include explicit references to the beauty of neutrons as a tool to explore the nano-world.[165] This is not an uncommon trope: the cover motif (a pyrite) for the 2016 Australian road map is described as "beautifully zoned."[166]

The image originally comes from the ESS organization and is one among many renderings that have been produced in the process, often provided by the architecture and design firms assigned the job of planning and building parts of the facility, manifesting a particular visual style. Nevertheless, it represents the challenges and dreams associated with something as large, political, and placed in the future as the European Spallation Source. These dreams are political, social, and individual. As sociologist and cultural historian Chandra Mukerji notes in her work on the Canal du Midi, infrastructure building can encapsulate "dreams on dreams on dreams,"[167] and through adopting this type of visual in a national document of considerable import, the Swedish Research Council demonstrates its engagement and complicity in such political, societal, and scientific dream making.

Visual representations of the type discussed in this section largely reinforce and strengthen the idea of infrastructure as detached technology-driven progression. Given the power of visuals for infrastructure making, it makes sense to consider not only how alternative framings of infrastructure can be represented visually but also how visualization and, more broadly, multimodal engagement can help situate and enact infrastructure—for example, through manifesting multiple perspectives, tensions, inconsistencies, feminist ideology, or contextual sociocultural data.[168] Indeed, visualization can be seen as an active tool for humanistic and humane capacity building, as discussed in chapter 3.

Interlude: Research Infrastructure Day 2018

Research infrastructure gets manifested in different ways, including through national convenings for different constituencies involved in research and research support. This interlude concerns such a convening that occurred on October 16, 2018, in Stockholm. Research Infrastructure Day was organized by the Swedish Research Council, whose government-assigned responsibilities include the long-term planning of access to research infrastructure in Sweden.

The goal of the day was to present the research infrastructure system and funded research infrastructures as well as to create opportunities for existing and potential infrastructures to prepare for future proposals.[169] The morning session was primarily informational, whereas the afternoon session directly targeted prospective applicants.[170] The following account relates to the first principal session.[171]

The event took place in one of the generic but upscale venues close to Stockholm Central Station: World Trade Center Stockholm. The space in question had a cinema-style setup with "very comfortable armchairs." There were provisions in the adjacent lounge area before and during the event, and lunch was also included.[172] There was a sizable crowd present there, and I recognized the head of a major funding agency and university leadership from several universities as well as some researchers and participants from the Swedish Research Council and other agencies.

The introduction dealt with the European research infrastructure (ESFRI) road map, which had just been released at the time, as well as how it connected to the national system for research infrastructure and how the evaluation and assessment of projects function. This segment was followed by four presentations of ESFRI infrastructures with Swedish leadership. The first of these infrastructures was the European Spallation Source, and the presentation was carried out by their head of communications in what turned out to be a very sleek business-style presentation (figure 1.10).

Figure 1.10
Presentation of the ESS at the Swedish Research Council Research Infrastructure Day 2018.

One of the slides informed us that the ESS vision is to build and operate "the world's most powerful neutron source" for "enabling scientific breakthroughs." The following three presentations were more dialed down and closer to the actual research. The chair of the Council for Research Infrastructures went on to talk about strategic recommendations. The morning session was closed by an expert from the Swedish Research Council, a Swedish representative on the European level, talking in quite some detail about the mechanisms and challenges of the research infrastructure system in Europe.

The program was executed efficiently, and it was certainly useful to learn more about the operation of research infrastructure and the large projects presented. A couple of things stayed with me though. One concern was the general inattention to the broader context of research infrastructure, as evidenced by the strategic recommendations presented by the chair of the Council for Research Infrastructures in the main session. He listed ten priorities, including the need for increased investments and more resources, improved infrastructure coordination, better career pathways for staff, and support for Swedish research and industry to be more involved in instrument and technology development. These priorities reflect and confirm the established framework of research infrastructure, which is not surprising or wrong, but it is noteworthy that there was nothing to address issues such as gender and environmental impact or differences in opportunities between different educational institutions. The whole event felt very managerial and privileged, again not very surprising, but it became abundantly clear in that room how much research infrastructure is about leadership, management, and control. Also, most of the researchers present seemed to be well established, often with an administrative or institutional slant to their work (much like myself).

There was a uniformity to the presentations, not just because they were about sanctioned research infrastructures—all part of a system or template—but also because the narratives and visuals given were so similar, almost as if they were stacked on top of each other. The normative tendency was striking and made more tangible through the homogeneity of the people present. Also, the seven speakers in the main session were all men.[173] This can perhaps seem like an oversight or something that is allegedly inevitable (because research infrastructure is a male domain, and hence finding women speakers is difficult), but it does carry meaning in at least two ways. First, the leadership of the Council for Research Infrastructures at the Swedish Research Council was male as well as the leadership of the infrastructures presented. Second, the all-male program points to a limited awareness of categories such as gender in the organization that has been given the task of managing research infrastructure in Sweden. Such an all-male program would likely not have occurred at any other council part of the Swedish Research Council and was not in accordance with the internal policy. The policy is supposed to ensure equality in meetings where the council is the organizer or the co-organizer, reflecting the council's awareness of the relevance of the distribution of speakers at seminars and conferences.[174]

> **The Swedish Research Council is well aware of the importance of gender equality in the distribution of speakers at seminars and conferences. An internal policy has been developed during the fall to guarantee gender equality in meetings where the Research Council functions as the organizer or co-organizer.**[175]

The disconnect becomes even more pronounced as one looks at the track record of the Council for Research Infrastructures, which has been slow to conform to the goals set by the Swedish Research Council centrally. For example, in September 2017, the Swedish Research Council published their decision to fund a total of twenty-eight infrastructure projects.[176] Only four of the infrastructure council's approved grants (14.3 percent) had a female applicant.[177] Compare that to the council of Natural and Engineering Sciences, often seen as the most male-dominated field, which had 21.5 percent successful main applicants who were women (out of 325 approved proposals).[178] By drawing attention to these numbers, I am not arguing that application decisions should necessarily be prescriptive in terms of gender but rather that the Swedish Research Council more generally needs to consider the status of infrastructure critically and constructively.

Infrastructure as a Refuge

In a seminal article on the need to consider "messy" facts as well as standards in the editorial practice of textual scholarship, Martha Nell Smith makes the following observation about humanities computing (which, over time, largely developed into today's digital humanities):

> **When I first started attending humanities computing conferences in the mid-1990s, I was struck by how many of the presentations remarked, either explicitly or implicitly, that concerns that had taken over so much academic work in literature—of gender, race, class, sexuality—were irrelevant to humanities computing. . . . Humanities computing seemed to offer a space free from all this messiness and a return to objective questions of representation.**[179]

This point was picked up in a provocative, polarizing, and productive discussion in 2015 by Adeline Koh and Roopika Risam, who suggested that digital humanities was a refuge from race, class, gender, sexuality, and disability.[180] In my own work in humanities and technology, I had observed that the driving force was often (but not always) instrumentalism, standards, interoperability, and technology, frequently at the expense of other perspectives, including critical, environmental, and disciplinary concerns.[181] It was as if technological tools were taken to lack epistemic inflection and that such inflection was not deployable in the making of tools and systems. Smith's critique is certainly warranted, but as Tara McPherson points out, Smith's own work also demonstrates how such technological spaces can be or become productively entangled with different kinds of critical engagement, including feminist theory.[182]

There is a similar technocratic and objectivistic sensibility to infrastructure policy and work. As this chapter has shown, infrastructure is a complex category that gives rise to many tensions, including being strangely disconnected from the world while being embedded in it ("Infrastructure shapes our lives," as a technology-centered UK National Infrastructure Plan states).[183] Infrastructure is a playground for futuristic technology-driven visions often set in a proximate future and can seem free from considerations that characterize most other civic and academic areas. The chameleonic nature of infrastructures allows them not to be political, gendered, environmental, grounded, or epistemological while simultaneously being all those things. In this sense, infrastructure can be a refuge, a disconnected place with its own logic. It is not always, however, and it does not have to be—herein lies an opportunity.

This disconnect can be observed on different levels, in the actual infrastructure, in the organization of infrastructure work, in the frameworks and ideas surrounding infrastructure, and in policy work. It is prevalent in research infrastructure policy for the humanities, for example. There are no references to gender or race (literally) in three major and influential reports on humanistic infrastructure published between 2006 and 2015, and there is no explicit interest in questions to do with environment and labor.[184] Furthermore, there is little, if any, obvious critical or theoretical considerations in these (and other similar) reports. The HASS (humanities, arts, and social science) platform suggested in the Australian Research Infrastructure Roadmap from 2016 (see figure 1.11) is seemingly disconnected from actual research and anything other than data, standards, interoperability, and access.[185]

In such documents, the proposed infrastructural models—for example, infrastructure modeled as "access to data and tools"—almost never deeply embed humanistic criticality, humanistic research questions, or the uncertainty, challenge, and resistance that are so central to the humanities endeavor. Even when the connection between research and infrastructure is referenced in a Swedish policy document, there is an implied distance between the two, exemplified by the need to "enable access to and facilitate correct use of the infrastructure," as if infrastructure is in a separate location from research.[186]

Elements	National Research Infrastructure response
Integrated and coordinated HASS platform	Explore integration of networks for corrdinated access to physical collections and digital materials enabling the digitisation of priority specimens across all collecting institutions. This could include the sharing of digitisation infrastructure and standardisation of best practice for processes and interoperability with international research infrastructures.
Harmonised platforms for Indigenous research	Explore integration of existing institutional level capabilities across a range of data platforms: AIATSIS, ATSIDA PARADISEC, and NCIG linked to wider platform for integration across all digital collections and portals.
Harmonised platforms for social sciences research	Explore integration of social sciences data from multiple sources together with tools for analysis and visualisation. This should be linked to the broader integrated and coordinated approach to a HASS platform. Maintain priority for AURIN.

Figure 1.11
Table from the National Research Infrastructure Roadmap, Australian Government. Credit: The Australian Government Department of Education, Skills and Employment.

At the level of individual universities, as evident from the Chalmers University of Technology website, research infrastructure is often identified as a central concern:

> **The concept of a research infrastructure includes everything from advanced real or virtual labs to large databases, computer capacity for large-scale calculations and magnificent research facilities. In short, everything necessary to conduct truly outstanding research.**[187]

In this short fragment, we find many of the features of infrastructure discussed in this chapter, including the framing of infrastructure as progress, definition by example, technologically driven visions, superlative infrastructure speak, and seeing infrastructure as foundational. Given that research infrastructure is identified as necessary to carry out high-quality research, it cannot be seen as something peripheral or specialized. It reflects and manifests the institution, demonstrated in the way the institution allocates resources and presents infrastructure. How the foundation for the university gets built matters and is an institutional responsibility.

Chalmers lists ten established facilities that fulfill the criteria for being a Chalmers infrastructure as well as two more facilities that had been added more recently (as of November 2020).[188] Facilities included the Onsala Space Observatory, Chalmers Maritime Simulators, and the Facility for Computational Systems Biology. For each facility, the name of the director was given. Of the ten facilities, all were led by men. Men also lead the two recently added facilities. Chalmers is a male-dominated technological university with 73 percent male professors in 2018, but this pattern is still striking.[189] It is as if there was not even any optics involved and no reason to think about the presentation or to include a passage on how the institution is working toward recruiting women and acknowledging female leadership.

This is a question not just of numbers but also of the quality of research and the long-term success of the institution. Chalmers made a major commitment to gender equality in 2018 through the Gender Initiative for Excellence (Genie), which is an important faculty-driven effort for the institution.[190] However, the reports produced by Genie to assess the state of affairs do not refer to infrastructure, which further support the notion that research infrastructure is its own place or that gender equality and other important perspectives often do not reach all the way to infrastructure.[191] It should be noted that it has been reportedly difficult in general to affect structural change through the project, which is perhaps not surprising.[192]

How can we explain the disconnectedness of infrastructure policy and work? There are several factors that seem pertinent, including the organizational separateness of infrastructure work, the technological framing of infrastructure policy, and an instrumental and rationalistic view of infrastructure. The mechanics of infrastructure work play an important role, since it includes and excludes perspectives. Furthermore, infrastructure is displaced through often being placed in the future or in an ideal past.

These factors also contribute to creating a distance between the material, systemic, and conceptual levels of infrastructure. However, more is at play here, which involves the elision of the humanistic and humane. Infrastructure policy and work remain a place where it is permissible not to engage with perspectives and responsibilities that permeate most other societal and academic sectors. The argument is not that this is deliberate negligence or malicious behavior but rather that there continues to be a lack of expectations, directives, and awareness that we expect to find elsewhere.

Importantly, changing infrastructures is not only about including or adding different categories and perspectives or removing a problematic parameter but also about allowing inflection to be a driving force. We often need to change the basic logic. Instead of designing projects where the bias of digital systems is "solved" through fixing the algorithm or adding another one, we must imagine systems driven by such inflection and the imaginaries that we believe are central to our common future. In this work, we can be helped by critical, value-driven, and process-driven approaches to design and making such as design justice and critical making.[193]

The futuristic ethos of many infrastructure projects, such as the Internet of Things and AI, is often problematic, as it takes us away from here and now as well as distracts us from considering how something such as AI maintains social inequity through its design and logic. Timnit Gebru and her colleagues argue that we "should be building machines that work for us" rather than adapting to a largely predetermined future.[194] One way of shifting possible futures is to "interrupt the scene"—in a sense, changing infrastructural frames and frameworks—as suggested by Louise Amoore in her work on machine learning—for example, through looking at algorithmic branching points as places for political decisions.[195] Such work is partly imaginary, and as we will see, imagination can be a crucial quality for making or remaking infrastructure. The hope and imagination potentially enabled by infrastructure are necessary for reshaping infrastructures, infrastructural frameworks, and our lifeworld.

2

The Infrastructural Humanities

If a lead actor for the story of infrastructure were to be selected, it would not be the humanities. Humanistic disciplines have been reluctant to engage with infrastructure policy, considered peripheral to research infrastructure, and are generally not seen as infrastructural or technological. Not normally pictured with hard hats, we do not tend to think of humanists as builders of material structures. Yet, I suggest that there are several reasons for considering a more central role for the field.

First, until recently, the primary role of the humanities in infrastructure has been not as an actor but rather as the writer of the story of infrastructure. A long history of important critical scholarship on technology and infrastructure exists, and infrastructure studies, science and technology studies, and history of technology inform current thinking. These critical perspectives on "infrastructure" have become more varied and nuanced as a range of humanities disciplines have shown growing interest in infrastructure and as the humanities have expanded into areas such as digital humanities, environmental humanities, urban humanities, disability studies, and queer studies.

Second, the outside position of the humanities prompts us to ask different questions in order to understand infrastructure. The humanities have another standpoint than the dominant tradition of infrastructure, steeped in engineering, science, and technology, and their negotiation of research infrastructure—including tensions, resistance, shifts, and alignment—tells an alternative story of the emergence and implementation of research infrastructure. This story shows the infrastructural template from the other side, so to speak, and highlights how the development of a more central (but still peripheral) infrastructural position for humanities has come at a price, both for the humanities and for infrastructure work. More generally, the negotiation of humanities infrastructure provides a useful case study for how infrastructural frameworks are made and can potentially be shifted.

Finally, we have naturalized a sense of infrastructure that is a product of modernity and rationalization. It mainly depends on engineering, technology, and growth, which have constituted a useful model in many ways although lacking in valency. For example, city infrastructure is lighting, electricity, and transport rather than lived experience, ideas of value, or social equity. It is systemic and institutionalized, rather than bottom-up and local. The humanistic and humane can help us shift our attention toward a richer notion of infrastructure as well as ways of collaboratively changing the underlying frameworks. One condition for moving in such a direction is for the humanistic to be simultaneously critical and constructive, which is why the way the humanities engage and negotiate infrastructure—invoking issues such control, reluctance, and agency—and shape their own academic infrastructure, for example through existing paradigms of humanities infrastructure, is so important.

Engaging Infrastructure

Humanists engage with infrastructure in different ways. Most readily, infrastructure is an instrument, tool, and everyday texture. Examples include offices, libraries, humanities centers, administrative systems, software platforms and tools, support services, and academic conferencing. Such engagement is material, conceptual, and organizational. Academics are embedded in these and other systems as well as in the university as an institution.

Some of these infrastructures are formally classified as research infrastructures, such as the DARIAH-EU infrastructure discussed in chapter 1. Other types of infrastructure in the academy are normally not subsumed under research infrastructure as a category but can still be seen as infrastructural. An example of such infrastructure is classrooms and other learning spaces. They support and condition many types of learning and research activities, and like many other infrastructures, they can be deconstructed in terms of their spatial, institutional, and conceptual organization.[1]

This deconstructive possibility also applies to university sites and buildings, both contemporarily and historically. The humanities have a larger architectural, organizational, and symbolic imprint on campuses than narratives about their peripheral position might seem to suggest, not disregarding the many grave challenges to the humanities in many places. This is not just a historical artifact as signature humanities (and arts) buildings are still being built, for example at the University of Warwick (2021), Clemson University (2022), and UW–Madison (projected 2026) to name a few.[2] One reason for this imprint, apart from the logic of current campus plans, is arguably that the humanities are meaningful, attract students, and offer symbolic power to leadership, funders, and outside donors (a power that also attracts oppositional forces). There can be no doubt humanistic knowledge is deeply connected to the history and conceptions of the university.[3] This relation is reified through physical structures and their location in some schools, often contributing to a sense of importance. At UCLA, the Dickson Plaza (sometimes called the central quad) includes three original buildings with a strong connection to the humanities: Royce Hall, the Powell Undergraduate Library, and the Humanities Building (Kaplan Hall). The quad is not just a beautiful and convenient spot but also an arrangement that reinforces the university's commitment to the humanities. As some UCLA humanists write, the (their) site constitutes the "historic and symbolic center of the university."[4] Compare another highly ranked university, the KTH Institute of Technology, where the (successful) humanities are located in a generic nondescript building at the very periphery of the campus, far away from the 1917 courtyard building, other early architecture, and new signature buildings.

The humanities are anchored in the university but also constitute an innovative force. University libraries provide an example. The humanities have a strong historical association with libraries (sometimes being housed in them), which are often at the center of university environments—think of the Low Library at Columbia, the Wren Library at Trinity College, University of Cambridge, or the Old Library at Trinity College Dublin.[5] This naturally close connection has led to many developments, more recently when libraries have repurposed space in the wake of digitization, removal of old-style media centers, and a shift toward more active library spaces. For example, digital humanities labs and centers, maker labs, library labs, and other functions have become part of libraries in processes that have often benefited, challenged, and developed both the libraries and the humanities. This type of spatial and conceptual rearrangement occurs in many domains and time periods. Consider the development of the seminar as an institution in eighteenth- and nineteenth-century Germany, which originated in disciplines such as philology and history.[6] It created a new space, often literally, for conversational and critical practice that was markedly different from the lecture hall and lecture-based teaching.[7] It was more intimate, less open, and strongly localized in terms of space.[8]

> **When the University of Strasbourg was re-established in 1872, after the German conquest of Alsace-Lorraine, seminars were introduced in all humanistic disciplines, each with its own rooms and a large seminar library.[9]**

The institutional seminar has a genealogy that includes meetings (and seminars) that took place in the professor's home and activities in learned societies but emerged as an important innovation for universities driven by the humanities.[10] This spreading innovation was institutional and ideational as well as architectural—what we may now call "infrastructural." It was also inflected—for example in terms of gender—just as the modern seminar is conditioned in different ways (see chapter 4). Historian Bonnie G. Smith describes the nineteenth-century seminar as "a masculine marketplace of knowledge," which came to have a substantial influence on the practices of knowledge work in history as a discipline.[11]

Infrastructure also serves as an object of inquiry in humanistic and social science research, providing historical, cultural, and critical perspectives on infrastructure. The distinguished history of studies on infrastructure, much of which can be subsumed under the rubric "infrastructure studies," is long-standing and continues to be an expanding field. Infrastructure studies, which is interlaced with the fields of science and technology studies and history of technology, is naturally concerned with infrastructure as an object of inquiry. This tradition has one origin in the study of large technological systems, such as electricity supply systems and transport systems, where the systemic nature of such systems was foregrounded as opposed to starting out from specific technologies.[12]

Through pioneering research, scholars of science and technology studies moved to consider infrastructure as sociotechnological, relational (what is infrastructure to one person or group can be work to another), and as a site of contestation and delegation, including what is handled technically and what is handled socially.[13] More recently, the field has expanded to become even more multifaceted, including work on repair, maintenance, platforms, knowledge work, sustainability, risk, and a myriad of areas such as financial and critical infrastructures.[14]

There has also been a surge of infrastructural interest in fields that have not traditionally engaged with infrastructure research—including English, critical theory, cinema studies, and anthropology—where infrastructure can be an object of inquiry and an interest as well as a perspective or critical framework. Humanists and interpretative social scientists study civic and academic infrastructures across a wide variety of topics and perspectives, including the racial inflection of search engines, the infrastructural construal of austerity, the informational regime of global supply chains, the problematics of smart green infrastructures, and the cultural politics of mass digitization.[15] There is generally a stronger disciplinary and content side (beyond infrastructure itself) to such work compared to infrastructure studies. As noted, "infrastructure" is a newer phenomenon in these disciplines. In the introduction to a collaborative 2018 article on anthropology and infrastructure, anthropologist Penny Harvey writes that infrastructure demands their attention while also noting that the concept did not have much currency for the discipline fifteen years ago.[16]

In cinema studies, there has similarly been increased interest in infrastructure. Data from one of the leading conferences—the Society for Cinema and Media Studies (SCMS) conference—show a clear development over the past few decades.[17] Between 2006 and 2010, there were only two paper titles and no session titles with "infrastructure" in them, whereas there were eleven paper titles and two session titles including the word between 2011 and 2015. Between 2016 and 2020, forty-eight paper titles and fourteen session titles referenced "infrastructure."[18] The increased presence of the term evidences not only the rise of infrastructure as an object of inquiry for scholars in this area (the material transport of information, digital platforms, etc.) but also the growth of infrastructure as a theoretical framework and perspective.[19]

As a perspective and framework, "infrastructure" offers considerable purchase in the humanities with regard to reimagining humanistic disciplines, interdisciplinary assemblages, and epistemic positions. This use draws on the penetration of infrastructure thinking, the instability of the term, and a turn toward material engagement in the humanities. As an example, the research group on "Reading for Infrastructure, Infrastructure for Reading" at the UC Humanities Research Institute explores infrastructure as a method and "theory-machine" to facilitate readings of racial relations, labor, and environment.[20] The metaphoricity and materiality of infrastructure are said to have the potential to facilitate "unexpected theoretical work."[21] Although the group does not focus on infrastructure as an object, their work sheds light on the construction of infrastructure, not least in their analysis of keywords such as ecology, exhaustion, and breathing.[22]

Infrastructure can thus become a method of interpretative analysis—a lens or a formation on a par with "network" for Internet-related scholarship.[23] For example, infrastructure as a method can realign the epistemology or disposition of media studies in the direction of paying more attention to distribution and the material aspects of media distribution.[24] Media scholar Lisa Parks argues that we need to consider what media are fundamentally made of, how they arrived as well as what they represent, and their history of style and meaning.[25] This infrastructural disposition, exemplified through Parks's illuminating study of mail sorters, power poles, and satellite dishes, foregrounds questions of distribution and resources, including bodily acts (e.g., tossing mail into bags) and material arrangements (the physical sorting system in the mailroom).[26] In a more general sense, infrastructure may also serve as an organizing principle and theoretical agenda for the humanities through conceiving infrastructure as the "operational experience of culture itself."[27] Media theorist John Durham Peters describes modernity as a "proliferation of infrastructures"[28] and goes on to introduce "infrastructuralism" as a new "ism" following structuralism and post-structuralism, drawing on the basic and boring as well as "all the mischievous work done behind the scenes."[29]

Infrastructure is given considerable theoretical power and transformative epistemic potential in these examples. This promised transformation is largely restricted to the humanities and their interpretative frameworks and hence our understanding of the world. While important intellectually and methodologically, this work does not normally seek to introduce ways in which humanistic engagement with infrastructure can be used to affect social and material change.

Another mode of infrastructural engagement in the humanities concerns infrastructure policy and, more generally, infrastructure as a framework. For example, when humanists negotiate, resist, accept, or devise models for research infrastructure in the humanities, they engage with infrastructure on a policy or framework level. Such engagement can be active as when trying to change or adapt infrastructure policy for the humanities (as well as actual infrastructures) or affecting how social media platforms handle moderation or bias in the underlying datasets.

Control and Power

Academics have different degrees of control over different infrastructures and engage with them differently. An administrative system, such as UCPath—which is used for payment and benefit transactions across several campuses in the University of California system—may seem close to impossible to change or tweak for an individual employee. It has been rolled out despite known problems, which, among other things, has led to employees not getting their paychecks and benefits on time.[30] These weaknesses affect the workforce negatively and unevenly and cannot be separated from the academic work.

Similarly, scholars are subject to the administrative systems used to solicit proposals for external funding. Submission processes normally leave little negotiation with regard to word counts, required categories, and imposed formats. It would be a mistake to regard such systems as purely administrative, however, as they condition and shape the articulation of scholarly ideas.[31] They also control the time cycles of academic work. These systems are hard to change for several reasons, including their administrative nature and the power of funding agencies, but this should not prevent attempts to revise them, ideally through collective efforts.

We can have more control over other types of infrastructure, including departmental space, the way we set up our work and teaching environment, database structures for research projects, or what events we choose to attend (given funding). These configurations affect our work and lives and have inflection. The way we build scholarly databases has epistemological consequences, the reliance on presentation software structures the way we teach, office (and lab) space manifests power, and there is an environmental footprint associated with conference and research traveling.[32] The level of control is contingent on multiple factors, including how privileged a person is in terms of hierarchy, roles, and support. A tenured professor is in a very different situation than a graduate student or a departmental administrator. Responsibility and complicity generally increase with privilege and power.

Changing Departmental Space

Imagine rethinking the spatial configuration of a university department as a means of making the department more just. Humanists would seem to have the insights that would motivate such a reconfiguring of space, as they study how space manifests and produces categories such as power, gender, and ideology.[33] It is rare, however, to see senior faculty object to the large windowed offices and other privileges that tend to come with higher status and seniority. There are a range of factors that play into this scenario, including individual preference and the need for space that may come with more senior positions or certain types of work. There can be no doubt, however, that there are structural patterns at play. For example, department administrators are less likely to have their own space, and faculty are more likely to have more space and more windows. In a social science department where I spent several years, the administrators had interior cubicles and faculty had windowed offices. In general, faculty spent little time at the department, whereas the administrators were there most of the time.

These infrastructures are heavy and difficult to change, as they inscribe power, tradition, and standards. Even when they are changed, they often retain or reinscribe the same set of values. Design historian Jennifer Kaufmann-Buhler argues that while the open-plan office is often presented as egalitarian, it reinforces the preferences and positions of the people in power and rarely includes the perspectives of service staff and lower-level support workers.[34] People attach personally to their professional space, some more than others, and there are many rationales for having or grabbing space. I have heard department heads rationalize their large offices through the need to have enough windows (or they would be seen as inferior) or the necessity to represent the institution well to important guests from the outside. I rationalized my oversized office at Umeå University through claiming that I needed space for a meeting table to be able to carry out meetings efficiently.

Change must be driven by heart and function and needs to be associated by a convincing narrative. The department in question could argue that the infrastructural plan was part of taking the department to the next level, meeting the requirements of the decades to come, and shifting the traditional hierarchies of academic work. The plan could be presented as a case study for the university as a whole—an experiment that would require central support as well as a license to circumvent some of the routines laid down by facilities management.

Such a plan—essentially an infrastructural one—could be framed in terms of function (staff tend to spend much more time at the department), ideology (as a department, we are committed to reinforcing hierarchies as little as possible, given the current structures), vision (staff–faculty collaboration is key to the success of the department, and we realize that we will increasingly be dependent on specialist staff), and opportunity (such a move might free up space for other types of infrastructure, including collaborative space, a podcast studio, or more space for graduate students). There would have to be broad support among staff and faculty, and the plan should ideally be heralded by powerful faculty and high-level administrators. The fight should not have to be mainly (or at all) taken by lower-level administrators or graduate students.

Enacting Values

A key challenge is to enact the values and criticality that are integral to humanistic and humane knowledge also in relation to our own lifeworld. Scholarship speaks for itself but can be amplified and extended through action and leadership. Responding to challenges through scholarship can create tensions, as when climate researchers are critiqued for frequent flying.[35] How do we relate to the Paris Agreement and the need to reduce carbon emissions drastically, not just as a critical or structural-level perspective but concretely in relation to our own business traveling?[36] Given that it is probably not going to be a question of flying or not flying for most academics, there must be serious consideration of what types of flying will be prioritized. What is the human and humane response? Should we prioritize early-career scholars? Should we discourage or stop participation in the large annual convention-like conferences that many disciplines host? Some of these issues—including climate change and low-carbon research—are addressed in a dialogue between two scholars in chapter 3 (interlude), while alternative models for academic conferences are explored in chapter 4.

It is important to be aware of our complicity, make informed decisions, and act when we can. Jane Ohlmeyer set an example when she, as director, made priorities about the space for the Long Room Hub at Trinity College Dublin. She chose to place management in an interior hallway on the bottom floor to give priority to early-career faculty:

> **Yet it's my final moments with Ohlmeyer that left the most lasting impression. Before I left, she whisked me up to her favourite spot in the building, the top floor where graduate and doctoral students carry out their research. "You know the architects wanted to put the directors and manager on the top floor, but I refused, because you guys coming behind, you are the scholars of the future. You are the lifeline we need to raise up. You are what's important."[37]**

Ohlmeyer took this decision from a position of power, which is one of the things that needs to happen. Consequently, the building's configuration reflects the vision of the Long Room Hub. Making the top floor a collaborative place for early-career scholars across departments is meaningful and speaks to valuing those that are the future of their fields. Such coming together of ideas and implementation outside infrastructural templates is rare and difficult. One barrier in the humanities is that we are often hesitant to engage with infrastructure in terms of designing and building.

Resistance

There are several reasons why humanists may not readily engage with infrastructure constructively. For one thing, humanists are often not comfortable with the strong technological and instrumental associations of infrastructure and do not necessarily think of themselves and their work as invested in technology and dependent on such infrastructure. This discomfort is strengthened and coproduced by infrastructure speak, which is largely non-inclusive in relation to the humanities, and tends to be technological, instrumental, futuristic, hyperbolic, and largely uncritical (chapter 1).

Second, the funding structure may also dissuade humanists. Most funding with a clear "infrastructure" label goes to other fields. Looking at the 2016 European research infrastructures road map, the allocation for humanities and social sciences landmark infrastructures (classified under social and cultural innovation) is about 1 percent for investments/construction/value and about 2 percent for annual running costs (compared to the overall costs).[38] Also, these infrastructures mostly encompass technology, tool building, and social science databases, more compatible with some parts of the humanities than others. Furthermore, it seems likely that most humanities-specific funding, especially disciplinary funding, would not go to proposals that are infrastructural in nature.

Third (as noted earlier), the humanities tend to locate infrastructure elsewhere, assigning it to the domains of science, technology, and engineering (academic infrastructure) or to the world outside academia (civic infrastructure) where it often becomes a distant object of critical inquiry. This critical interest rarely extends to the humanities themselves, however, and many humanistic disciplines do not readily engage with their own current modes of knowledge production critically, which, in turn, has contributed to inscribing the humanities as a place without infrastructure.

Additionally, the humanities are seemingly locked into a construal of themselves as institutionally underprivileged and threatened (not necessarily without good reason), which is not really compatible with acknowledging existing infrastructure. If the humanities are framed as a poor, threatened, and reactive enterprise, this notion is not well served by examining the large amounts of funding allocated to the humanities for space, technology, libraries, and various types of centers and institutes. These different parts make up an infrastructural whole, however challenged and deficient. Yet another factor is how humanistic and humane knowledge is viewed and engaged outside the humanities and the academy. The actors we typically see as the builders of infrastructure—including engineers, city governments, finance, and tech companies—are not used to involve humanistic and humane knowledge in any substantial way outside what may already be embedded in their disciplines or enter through students. The humanities may also be reluctant to engage with such actors or be comfortable with an auxiliary, detached, or even ornamental role. In chapter 3, we will consider such negotiation further with a focus on moving in a more engaged direction. Such modeling, however, must relate to what infrastructural and epistemological models are already encoded.

Models of Humanistic Infrastructure

Infrastructure for and with the humanities is shaped by a variety of factors, including scholarly and educational needs, conceptions about humanities work and the humanities, the dominant STEM-based model of research infrastructure, and higher education policy. Models of humanistic research infrastructure in policy and infrastructure work emerge as a result of these and other factors. In the following, I propose five analytical models: the solitary scholar model; the cultural heritage model; the digital cultural heritage model; the data model; and the communication model.[39]

These models—however overlapping and imprecise—can shed light on how ideas, policy, and implementation associated with infrastructure come together. Such knowledge can, in turn, give us insights about key assumptions and path dependencies that influence how infrastructure, particularly research infrastructure, gets framed and implemented. I argue that the way we construct ourselves infrastructurally inscribes what the humanities are and could be, which is why we need to be engaged actively in model building as well as in infrastructure building.

The Solitary Scholar Model: No or Limited Infrastructure

When Edward Ayers writes that the humanities "remain largely a solitary craft, inexpensive and undemanding in the larger institutional context," he presents a particular (and common) framing of the humanities.[40] At a conference on humanities centers in the US, Daniel Callahan said that one of the attractions of the humanities is that it seems to offer to take humanists "out of the world" to the realm of books and ideas.[41] He also added that humanists often like to work alone, not in groups. Art historian Murtha Baca describes challenges associated with digital art history, one of which is the "tradition of the scholar toiling in isolation, which has prevailed in our profession from the beginning."[42] The notion that the humanities is largely a solitary endeavor with little or no infrastructure is common.[43]

While few people would probably subscribe to an extreme version of this model of infrastructure—which I call the "solitary scholar model"—it is commonly confirmed in informal conversations with humanists, either as something we need to challenge, given the shifting conditions of knowledge production and the state of the world, or as something we need to preserve as a space for contemplation and removal from the increasing pressures of a neoliberal society and academy.

Newer fields within the humanities often use this framing to emphasize how they embody change, especially in relation to moving from the solitary scholar model to a team-based one. For example, electronic collaboration is contrasted with the "paradigm of the lone humanities scholar whose work focused on a single period or text or person" and digital humanities is framed as a "fundamental shift in the paradigm of humanities, from the model of the lone scholar to that of the interconnected team."[44] These descriptions set up simple contrasts for effect, but they also manifest and reinforce diverging ideologies.

The product of the solitary work of the humanities is typically taken to be the monograph:

> **Collaboration has not traditionally been taught to humanists, whose normal working practice has been to produce a monograph alone.**[45]

The infrastructure required to produce books is not seen as collaborative or even infrastructural, whereas the team-based collaborative model is readily associated with technology and infrastructure. Contributing to this limited perspective is a sense that the humanities do not have adequate knowledge or training for infrastructural work. For example, a 2006 European-level report on research infrastructure declares that the humanities and social sciences are weak in terms of their "culture of standards."[46] There is also typically little attention paid to what the critical work of the humanities—including research on the role of standards culturally and historically—could add to the shaping of infrastructure or infrastructure policy.[47]

The solitary scholar model can be contested in at least two ways. First, almost all work in the humanities is collaborative one way or another, although the output—conventionally monographs—often conceals the collaboration playing out in seminar rooms, at conferences, in classrooms, and through the various ways texts in progress are shared, discussed, and reviewed. It may be true that the humanities at large privilege individual work and are less technological than some other fields, but this picture has been changing and has never been uniform. Second, there is no necessary correlation between solitary scholarship and a lack of infrastructure or low cost. John Durham Peters argues that the humanities have media and material conditions as much as other forms of inquiry.[48] Even the humanist alone in their study—typified by Renaissance painter Antonello da Messina in the master work *Saint Jerome in His Study*, depicting an early ideal of solitude and learnedness—is clearly infrastructurally embedded (figure 2.1).[49] Saint Jerome's work is reliant on a desk, various tools, shelves with materials, and an extensive multiple-tier space that has symbolic and material qualities.[50]

An anecdotal example of the persistence of the solitary scholar model of infrastructure comes from a Swedish Facebook discussion in 2017 where the cost of massive science infrastructure was discussed.[51] The author of the original post confirmed that there are indeed major costs for infrastructures, such as the European Spallation Source, but there are also significant costs associated with humanistic infrastructures, such as the National Library (Kungliga biblioteket) and the National Archives (Riksarkivet). One commentator, a senior humanities professor, stressed the point that the costs of humanistic infrastructure are insignificant ("crumbs") in comparison.

However, the annual running costs for the National Library and National Archives in Sweden (about SEK 800 million) at this time were a bit more than half of the estimated running costs of the ESS (SEK 1,400 million). While these cultural heritage institutions serve many functions and constituencies,

Figure 2.1
Saint Jerome in His Study, circa 1475. Credit: Wikimedia Commons.

there can be no doubt infrastructures for the humanities exist and that their historical establishment was a massive long-term investment on a par with modern science infrastructures.

As this anecdotal example shows, there is a tendency for the humanities to downplay the importance and cost of their existing infrastructure, which can be ascribed to the sense that the humanities do not have any or much infrastructure. Not being seen as infrastructural can sometimes be strategically advantageous. In another informal context, I heard a leading representative for the humanities at a prestigious Swedish university explain how they deliberately did not frame a new initiative as infrastructure, as they were tapping other funding and management streams. This strategy presumably made it possible for them to maintain a largely solitary scholar model and avoid the research infrastructure game. On the other hand, an infrastructure-framed solution may have forced different modes of engagement.

Generally speaking, it would seem advantageous for the humanities to own the question of infrastructure, old and new, and acknowledge actual epistemologies, old and new, which means moving away from the solitary scholar model (but not necessarily individually based work).

The Cultural Heritage Model: Cultural Heritage Institutions and Materials as Infrastructure

The cultural heritage sector (archives, libraries, museums, and galleries) makes up another model of humanistic infrastructure, partly overlapping with the solitary scholar model. These institutions, particularly libraries and archives, have played a central role in the infrastructural imaginary relating to the humanities. They have not only provided materials for humanistic work but also been key places and destinations for humanists. Archives are described as the territorialization of curiosity by ethnology scholar Orvar Löfgren, thus invoking a powerful frame that would seem useful for humanistic infrastructure work.[52] He also describes how his first professor in history asked the students to always be "naked" in the archive and makes a parallel to disinfection fluids and the white coats in science and medicine.[53] This comparison, while intriguing in its own right, also points to how science often becomes a measure stick and framing strategy for humanistic infrastructure. It also reminds us that institutions such as archives have inflection—for example, they are gendered and fetishized.[54]

Similarly, as a place, the library seems to manifest humanistic work and pride. Humanists interviewed in anticipation of a possible move of the Humanities Library at MIT in the early 2000s are clear about the value of the library beyond the collections.[55] In the report, they talk about the symbolic importance of the library as a space, how it is connected to their identity as humanists, and how it is an oasis that gives them dignity and academic worth. They also compared it to other types of spaces and resources on campus:

The library is our lab.[56]

Comparing the library to the lab can be seen as an infrastructural move, as labs would readily be seen as infrastructures. The material qualities of the library are emphasized in the documentation from MIT, including the network ports, the large windows and natural light, and the location on campus, strengthening the connection to infrastructure.

Cultural heritage institutions have a long history, which greatly predates "infrastructure" as an idea and framework. It is sometimes used to anchor humanities research infrastructure policy, for example through invoking the Mouseion in Alexandria.[57] The formal infrastructuralization of these institutions, however, is a recent phenomenon, and partly connected to the digitization of cultural heritage and shifting conditions of knowledge production and representation. Another influence is the growth of research infrastructure as a framework and the increased emphasis on knowledge as a commodity. Institutions such as libraries have come to be seen as infrastructures, networks of infrastructures, or platforms.[58] This development has been accelerated by a need for libraries to rethink their space and resources, given increased digitization and outside pressures. As noted earlier, this means that there has literally been space to consider and implement new humanistic infrastructure-related initiatives within libraries, including digital humanities centers and maker labs.

Change in these institutions is not just here and now. While the National Library and National Archives in Sweden date back to the beginning of the seventeenth century, they were established as national and independent resources in the 1870s. Motivated by a national need to collect, preserve, describe, and make materials accessible in a coordinated way, the changes were part of a move toward national cultural heritage institutions (free of the church and crown) that can be traced back to the late eighteenth century.[59]

Figure 2.2
The National Library in Stockholm. Photo credit: The National Library of Sweden.

When the new National Library in Humlegården in central Stockholm was opened to the public in 1878 (figure 2.2), it had been designed for two hundred thousand books and ten employees, which soon proved to be not enough.[60] The building manifests this institutional and infrastructural expansion, much in the same way as the ESS campus does for the European Spallation Source.

While the National Library is nowadays framed in terms of research infrastructure in the ordinance from the government, this was not the case before 2018, and cultural heritage institutions are generally not part of the logic of research infrastructure policy.[61] For example, there is no infrastructure road map that includes the Swedish National Library and the National Archives. Significantly, there does not appear to be a perceived end date to these institutions. The European Spallation Source may come across as more or less permanent, with a budget calculation that extends until 2065, but it is still transient in comparison to the National Library or the Nordic Museum, which do not operate on a timescale and seem to exist outside the logic of road maps.

These institutions typically have multiple stakeholders and roles and have a different funding stream from higher education infrastructure. They also have their own agendas and traditions and tend to have a strong position in terms of policy, especially on national and pan-national levels. Often, several ministries or government bodies are involved for cultural heritage and research, which makes coherent infrastructure work more difficult. And because galleries, libraries, archives, and museums within higher education tend to have dedicated funding, there is little incentive for the humanities to reclassify these as research infrastructure. At the same time, digitization of materials and the increased reliance on digital tools have opened up an area that is easier to see as research infrastructure: digital cultural heritage.

The Digital Cultural Heritage Model: Giving Access Through Digitizing Everything

Digital cultural heritage provides a pervasive infrastructural model. This infrastructure brings together the interest of cultural heritage institutions (especially libraries and archives) to collect cultural heritage and make their collections available; a technologically driven interest in metadata, access, and standardization in digital humanities and related disciplines; and the technological and data-driven framing of research infrastructure. It also draws on the long history of cultural heritage institutions.

According to the American Council of Learned Societies (ACLS) report "Our Cultural Commonwealth," a key objective is to create "universal access to knowledge anywhere and everywhere."[62] The mission is predicated on the idea that studying human cultures and creativity is based on "access to the records of the past."[63] Digitizing materials and creating repositories become a goal in itself, which often seems more important than carefully considering if and how the material will actually be used. Medievalist and digital humanist Jeffrey Schnapp argues that historical materials need to be animated by users or they may never come to life:

> **Few guests show up at the party because digitization in and of itself doesn't resolve questions of access, visibility and availability for actual use.[64]**

There is a belief expressed in the ACLS report and in many other policy reports that making cultural heritage materials available and linkable will lead to scholarly breakthroughs and societal impact. A European-level policy document on research infrastructure states that making cultural heritage accessible and linkable "opens a new frontier for Humanities research for addressing 'grand challenges.'"[65] However, the connection between accessible cultural heritage and addressing grand challenges is not explicated.

In a 2021 newspaper opinion piece, the heads of five Swedish cultural heritage institutions argued for the need to digitize Swedish cultural heritage, especially given that only a small part of the collections is currently available digitally.[66] Nothing in the article references anything in terms of action beyond digitization and access: there is no consideration of *how* the material will be activated and used. Just as with the ACLS report, however, the potential gains listed are substantial. The authors stress that digitizing cultural heritage is an issue of democracy, openness, and access. They also assert that everyone, irrespective of background, should be able to access the cultural heritage that has formed and continues to form us. Furthermore, the writers reference the benefits for research, innovation, and the national economy. In all these examples, considerable agency is given to the infrastructure itself, framed as digitization and access.

Digitizing archives and other materials has obvious merits, but it is also a dream. Digitization and media researcher Nanna Bonde Thylstrup argues that mass digitization is not just a rational project but one infused with desire, nostalgia, and power.[67] As we saw in chapter 1, the infrastructural and technological imaginary can be a strong driving force, which sometimes functions as a surface that obscures or elides the underlying complexity. The argument is not that cultural heritage institutions are unaware of these complexities but rather that the infrastructural dream pushes development in a certain direction. The infrastructural model of digitizing everything is mainly a matter of mass digitization and "traditional non-selective practices" rather than critical digitization.[68] Moreover, by making "digitize all cultural heritage" into a frame—as if it was actually possible—priorities made along the way are backgrounded. What gets digitized first and why? And how are these priorities related to projected gains—for example, supporting our democracy or facilitating high-quality research?

What gets digitized tends to be established and canonized cultural heritage. One relevant question is exactly what cultural heritage will be made available to everyone, "irrespective of background"? Cultural heritage institutions are built around collections that reflect institutional, epistemological, and societal conditions and biases. For example, it is likely that marginalized groups will find their cultural heritage underrepresented. What is absent from the collections or silent in them is also important.[69] These gaps relate not only to power, identity, gender, and racial issues—crucial as they are—but also to whole swathes of material, including "movement heritage" and the vast materials produced by an increasingly digital society.[70] It can also be argued that there is a media bias within cultural heritage institutions, for instance in relation to mass media.[71]

Digital cultural heritage as an infrastructural model aligns with the epistemology of cultural heritage institutions (particularly libraries and archives) as they are traditionally configured. New digital infrastructure is often built on top of the old one. The ACLS report refers to collections in libraries, archives, and museums; some associated tools (such as bibliographies, concordances, and classification systems); and expertise (e.g., librarians, curators, and archivists). They claim that these and other elements have "extensions or analogues in cyberinfrastructure."[72] While making the digital into a replica of the analogue is a natural and common tendency, it often translates into highly conservative systems. Digitization may increase the speed and distribution of these once-analogue systems, but it does not necessarily change the core logic. The classification systems, tools, and collections remain roughly the same. This is why recent developments, including library labs and the expansion of community archives, can play an important role—if supported appropriately—in shifting the logic.[73] A powerful example of such work is the use of "bundles" as a conceptual-material tool to bring different types of data together in the project "SpeakingUp—Making Spoken Cultural Heritage Accessible for Research."[74]

Why access is not enough

Through an exclusive focus on digitization, access, and the grand scope of "digitizing everything" (which is a massive, costly effort), we risk downplaying related intellectual and ethical questions and paying too little attention to how materials are shaped through the way they are digitized and made available. We know that categorizations, standards, and representations come with predispositions and are laden in different ways. We also know that digitization and access may not always be appropriate and that the "how" is crucial.[75] Generic access platforms also typically assume a certain neutrality of the material, and limited effort is devoted to developing alternative platforms for working with and interpreting the material.

In a discussion of visualization, Johanna Drucker argues that it is impossible to maintain externalized user-independent notions of knowledge in relation to the critical approach of the humanities.[76] Hence, the epistemological impact—and potential—of the systems we build need to be an important concern. There is no such thing as simple and pure access, and the systems we construct to facilitate "access" can be thought of as intellectual middleware that structures and conditions the relation between the material and the interpretation.[77] For example, Jewish Studies scholar Todd Presner's work on the ethics of the algorithm and the construction of "atrocities" in the Shoah Archive and Visual History Archive demonstrates how certain narrative elements and content are privileged through the ontologies and metadata embedded in the system and through the material specifics of the interface.[78] This is not just a critical perspective but also about how we design tools and platforms meaningfully. In his study, Presner suggests that information architecture and databases (infrastructure) should not be considered apart from the subjective and interpretative practices of the humanities.[79]

If building infrastructure is also about critiquing and imagining infrastructure, there would seem to be a place for a more generative and forward-looking vision for cultural heritage-related infrastructure than digitizing everything, even if gestures are often made to what benefit will be created through digitization and access. Such projected benefits—whether it be stronger democracy, new research frontiers, or mobilizing the humanities to respond to societal challenges—are not separate from the material; require integration, inflection, and embeddedness; and cannot wholly be left to others.

Infrastructure work is ultimately also a question of responsibility and sustainability, as Johanna Drucker points out in the description of a 2020 UCLA event on the future of special collection libraries:

> **These are fundamental ethical issues [that] include the tension between preservation and use of historic sites and grounds, the need to create viable pipelines for future generations of librarians to provide leadership, the high ethical and ecological costs of digitization, the threats posed by inequities in labor and work within the profession, and the possibilities of considering de-accessioning, once a completely taboo topic. In essence, every aspect of our work needs to be considered within the terms of sustainable activity.[80]**

What Drucker suggests is an expanded notion of sustainability for cultural heritage institutions. This is an opportunity, but it needs to be acted upon. For example, one must wonder why the 2019–2020 annual reports from the

National Library and National Archives in Sweden do not mention the words "sustainability," "environment," "ecology," or "climate" (apart from the sense of climate control of collections) when we know that digital storage and processing have profound energy impact.[81] This trend is continued in the reports from 2021 and 2022, although there is some limited environmental engagement from the National Library (especially in the 2021 report).[82] This apparent lack of attention can also be noted in a 2023–2025 plan for the National Library, where there is basically no emphasis on sustainability and the environment.[83] All in all, based on the reports, there is very limited engagement with these issues and certainly no proactive or directional efforts.[84] However, digitization and digital infrastructure cannot exist in isolation from long-term ecological concerns and climate change, the ethical consequences of making heritage available, or, in general, the state of the world.

The Data Model: Data as Infrastructure

A related model is to see humanistic infrastructure as databases, registers, or simply data. Data structures are commonly referenced when framing the work of the humanities and social sciences in infrastructure policy documents. A data model approach to infrastructure also encompasses tools used to manage and analyze data and can be linked to open science and open access as a movement, for which sharing data is a central concept.

Databases and data as synchrotron light and gateway

In an early infrastructure consultation with social science and humanities leadership across Canada (2009), research databases were described as the "'synchrotron light source' of the SSH."[85] This quotation not only reflects the need at the time to make a policy case for social science and humanities infrastructure but also relates to the results of the consultation, where databases were the most common request ("In almost all submissions, support for databases of various types is the prime focus of response").[86] Ten years later, the Swedish Research Council listed the following key examples of humanities and social science infrastructures in an overview: large registers and databases, digitization of collections, and access to laboratories for research (in that order).[87] In the subsequent discussion, they give registers and register-based research particular attention, as well as the need to link register data. As with many other similar documents, both these sources emphasize the linking and interoperability of data. Interoperability is a common keyword often taken at face value—as a principle or intrinsic trait—although achieving interoperability is typically a challenging task on several levels, even in single datasets that may seem homogenous.[88] Some types of data, such as those from the Global South, are more likely to resist homogenization in frameworks and tools primarily created for a Westernized, Anglophone context, potentially becoming invisible.[89] At the same time, as Annalisa Pelizza observes, infrastructure studies tends to make interoperability and integration into a definitional property of infrastructure.[90]

Data become a category in their own right. Interoperability presupposes data to be compatible, which means that there is a drive toward standardization and unification. As noted earlier, it is sometimes observed that the humanities are lagging behind in relation to social sciences, and the Swedish Research overview points to the fragmentation of the humanities as a challenge.[91] More generally, data are taken to be necessary for the research process, much in the same way as modeled in science-based framings of research infrastructure.

> **Within the social sciences as well as for the humanities collection of data is essential and could be referred to as the fuel for the research process.[92]**

While data are given a central role and considerable agency in this excerpt from an ESFRI report, referring to the "humanities collection of data" invokes the practice of collecting material, thus giving more humanities (researcher) agency within the epistemic frame established, balancing it somewhat. In other cases, data are given even more agency, as with "data-driven research."[93] This type of framing can be seen in funding calls such as the Digging into Data Challenge (funding agencies in the US, the UK, and Canada) and data-driven research (several Swedish funding agencies).[94]

The data model has several origins and contingencies. Partly, it operates as an extension of the digital cultural heritage model. The step from collections to databases is small. Furthermore, databases are common in the history of humanities computing and digital humanities as well as part of a longer history of archives, collections, and corpora. Digital databases have been perceived as a need within the humanities more broadly for quite some time, as we saw in the Canadian survey referenced earlier.

The adoption of data systems has been heavily influenced by the social sciences, where databases, registers, and surveys have a central infrastructural and methodological role. For example, the three social science infrastructures listed in the 2018 European ESFRI road map are all survey data infrastructures or digital archives.[95] Because the humanities and social sciences are often grouped together, not least in infrastructure policy, the category of databases can easily be extended to the humanities, becoming an infrastructural cornerstone for both.

There is also another important reason why the databases and data model has been powerful in humanistic policy work. Databases and, more generally, data align well with the conventional framing of research infrastructure. Put differently, the STEM-based model of research infrastructure maps well onto a data-centric vision of humanistic infrastructure. It is therefore not surprising that infrastructure thinking in the humanities has tended to develop in this direction. Data have served as a gateway for the humanities and social sciences into research infrastructure policy.

Aligning to data

A useful historical example is the ACLS report from 2006, in which humanistic infrastructure is clearly modeled on a STEM model, particularly the Atkins NSF report from 2003.[96] The ACLS report argues that humanities scholars and social scientists "will require similar facilities but, obviously, not exactly the same ones."[97] The report adds that the humanities and social sciences will have different digital objects than the sciences and that software toolkits should be shared across the natural sciences, social sciences, and humanities. However, sharing should continue "only up to the point where the nature of the data begins to shape the nature of the tools."[98]

The mapping suggested in the ACLS report seems problematic.[99] The model is driven by tools and technology, closely mapped onto the NSF template without any noticeable resistance. Many humanists would hesitate even to talk about their material as "data" or "digital objects." The report also makes a sharp delineation between the digital objects (data) and tools. Tools are expected to be highly generic, and the point at which the data start to shape the tools occurs late. This model does not seem to make provision for tools that are informed by the data and specific research questions or driven by critical epistemologies, whether theoretical or methodological. It is clear where the epistemological starting point is, just as when James Smithies suggests that to create successful digital tools, we—despite humanistic resistance to "technical work *and the epistemological positions that enable it*" (original emphasis)—need to employ positivist and utilitarian approaches "to some degree and with necessary modifications."[100] While there is some truth to this position, I argue that "necessary modifications" and "similar facilities" are not generally sufficient for challenging the standard epistemic framing and epistemic commitments. Such framings and commitments are of many different types and not just linked to a technological imperative.

Data are not just out there, and they are not the same thing to different disciplines. An infrastructural vision built around data will be more compatible with some humanistic disciplines than others, notably areas that see themselves as collecting and processing data computationally, including computational linguistics, environmental archaeology, and text analysis. Such fields have a language and readiness that make it easier for them to align with data as an epistemic commitment. It is true that digitization means that more and more disciplines are engaging with materials that could potentially be described as data, but even so, the reach of the data as infrastructure model is limited, and it also tends to exclude disciplines (and perspectives) that shed critical light on concepts such as "data."

The European Open Science Cloud (EOSC), launched in 2018, can be used to exemplify some of the points made above. EOSC is a European-level effort to facilitate storage, management, and reuse of research data across disciplines and countries.[101] Data are central to the initiative, and EOSC can be described as a data-based infrastructural entity. Identified as part of the digital single market strategy by the European Commission, it is clearly embedded in the wider European project.[102] The EOSC is conceived of not as one infrastructure but rather as an initiative that includes infrastructures and federates data and connects to the European-level (ESFRI) infrastructures. Generally, open science organizations such as the EOSC align well with and reinforce data as a model for infrastructure. The EOSC sets out to cover all disciplines, but since it represents an epistemology driven by data as an idea, it will be more compatible with some disciplines and areas than others.

It is not surprising that the first humanities and social sciences project under the EOSC pilot phase was a tool to facilitate collaborative semantic enrichment of text-based datasets.[103] The broader rationalization provided for this tool identifies social sciences and humanities as facing a fragmented research landscape, where EOSC can assist the community through helping to structure and integrate ESFRI infrastructures (such as CLARIN and DARIAH) and digital humanities organizations.[104] EOSC becomes a super structure centered around data and already existing infrastructures and organizations.

Data can, no doubt, be a useful category for imagining and implementing humanistic research infrastructure. It is absolutely central, though, that the humanities attend to their epistemic and critical requirements in infrastructure building and also consider how qualities such as uncertainty and asymmetry can be part of resultant infrastructures. This knowledge is relevant to research infrastructures more generally, as the humanities can help challenge and shift technology-driven infrastructural agendas. In order for this to happen, however, the humanities must not leave their epistemic and critical inflection behind when engaging with infrastructure. Humanistic infrastructure work needs to engage data meaningfully and pragmatically, however, which means that the humanities cannot stand outside the larger infrastructural ecology or the computational structures in place. In many cases, it is reasonable to align closely with infrastructural templates and standards, but there is also more to humanistic engagement with data and infrastructure.

It is the job of the humanities to engage data critically. "Data" are not just an infrastructural category, algorithmic input, or scholarly materials but also a critical category. There are no global climate data without processing through data models.[105] The collection of data over time tends to be local, contextual, and delicate.[106] Database formats—including hierarchical, relational, and object oriented ones—reflect social and organizational developments.[107] Power, privilege, labor, and oppression are built into datasets, databases, and algorithms.[108] Epistemological positions are embedded in data representations.[109] If making infrastructure is about remaking worlds, as the Australian policy report states, we need to consider how these and other perspectives matter for infrastructure policy and for how we frame and design data-rich infrastructures.[110] In chapter 3—pushing this point further—I argue that we need to let critical inflection be one driving force or logic for designing infrastructure.

The Communication Model: Collaboration, Communication, and Publishing as Infrastructure

Although there is reference to communication in some early research infrastructure policy documents, as when the Canada Foundation for Innovation was created, it did not translate to communication-centered models for humanistic infrastructure.[111] Over time, however, a renewed interest in communication, collaboration, and publishing ("circulation")[112] as infrastructure has emerged, alongside research on scholarly communication infrastructure, the expansion of public humanities, and the changing landscape of scholarly publication.[113]

In the light of reduced funding for Swedish national journals in 2013, humanists compared such publications to large science infrastructures, making an argument for supporting humanistic publications.[114] Academic presses, which have been under considerable pressure, have also turned to infrastructure as a way of reframing their operation and arguing for increased resources. In 2021, two directors of North American university presses wrote about the need to reenvision infrastructures for the humanities.[115] In their opinion piece, the directors assert that university presses are "mission-critical infrastructure" for humanities research. They call on institutions of higher education and funders to act, and again a comparison to the sciences is made:

> **These stakeholders have long seen the value in investing significant resources to sustain science infrastructures that contribute to a common good. It is essential to their mission that they collaborate and invest with that same care in the crucial infrastructure for humanities research embodied by the network of university presses.[116]**

In contrast to the data model of infrastructure, the framing of university presses as infrastructure is more strongly grounded in the institutional humanities. Academic presses have been central to much of the humanities for a long time, and they tend to be close to the groundwork of the disciplines. It could be argued, perhaps, that the vision presented still reflects a conventional view of publishing and the humanities, although reference is made to a range of projects and perspectives that are progressive and point forward.

Digital humanist and English scholar Kathleen Fitzpatrick argues that it is not the academic book itself that has stopped functioning but rather the system associated with its production and dissemination.[117] An infrastructural response to this challenge seems right, but it would take upending the current system in a concerted effort and convincingly relating to an ecology of infrastructures. An additional challenge is that university presses as an infrastructural vision cannot easily do the job of the humanities, despite them being grounded in the humanities. Academic publishers are not the humanities and cannot stand in for the humanities, which themselves are changing.

The university press–driven vision of humanistic infrastructure is partly induced by outside pressure and a need to adapt and change. The response, however, does not necessarily go outside the institutional frame that is being challenged, at least not in a systemic manner. Related developments in communication, public engagement, and outreach can shed light on a humanistic infrastructural imagination as a response to these challenges.

Challenges to a public framing

When the Trump Administration proposed to eliminate the National Endowment for the Humanities (NEH) in the US in March 2017, the NEH response was partly to frame the humanities as public outreach.[118] The statement below (dated March 16) came from the chairman of the NEH and was also followed by a similar, somewhat longer statement by the NEH (dated May 23) just after the administration had eventually granted funding for 2018.

> **These grants have reached into every part of the country and provided humanities programs and experiences that benefit all of our citizens. Residents in Whitesburg, Kentucky are preserving the photographs and films of their local Appalachian region through Appalshop cultural center. Veterans returning from war in Iraq and Afghanistan connect with classic texts and the public through Aquila Theatre. Students, teachers and historians have access to the papers of Founding Father George Washington. Through these projects and thousands of others, the National Endowment for the Humanities has inspired and supported what is best in America.**[119]

This is an important statement produced under very difficult circumstances (the NEH chairman later chose to step down). It is not surprising that the framing is one of public engagement and outreach, which is an important mission of the institution, but it also reflects political pressure (over time), a general pressure to demonstrate usefulness, and a desire to counteract any sense that the NEH supports peculiar and leftist ("dangerous") scholarship.[120] There is nothing wrong with this strategy, and since the NEH was not eliminated, it was a successful one. Nevertheless, by construing the humanities as mostly about outreach and public engagement, other perspectives and results get backgrounded.

For example, in a concurrent NEH strategy document, the first objective listed is to facilitate "basic research and original scholarship in the humanities," and elsewhere, the NEH's website lists several thousand books as achievements of the organization.[121] While the emphasis on outreach at the expense of research by the NEH chairman in the 2018 address was understandable under the circumstances, it is worth considering carefully—not least in relation to modeling humanistic infrastructure—as it connects to a larger context of instrumentalism, focus on communication, and politically charged challenges to humanistic "edginess."

While the humanities have a long and important history of engagement outside the academy, it could be argued that the renewed interest in "public humanities," as exemplified in the NEH statement, has shifted the framing of the humanities, especially in the US. The institutional framing of public humanities goes back to the 1970s, in the US at least, and has become an important part of institutional humanities.[122] It is a framing that is not unproblematic. It can be argued that with this model, we risk separating scholarship and outward-facing activities,[123] paying more attention to communication than scholarship, centering academics and educational institutions rather than the outside world,[124] and framing the humanities as a simple outward movement from the academy into the community.[125] "It should be noted, however, that there are upcoming initiatives—such as those labeled "public scholarship"—that model outward engagement in a more complex and reciprocal way, although these often do not attract as much infrastructural attention.[126]

A public framing of the humanities (humanities as public engagement, outreach, communication, etc.) can be limiting. In infrastructural terms, a public framing of the humanities, just like the digital cultural heritage model, highlights only specific aspects of the humanities, which then become a model of humanities infrastructure and, in some ways, a model of the humanities.

A critique of collaboration and exchange as humanistic infrastructure

An articulation of this model comes in the form of a report on new knowledge infrastructures and the future of humanities and social science collaboration published by three Swedish funding agencies and knowledge institutes in 2016.[127] In many ways, the report is a progressive and useful document that relates to the role of the humanities and the current emphasis on collaboration and knowledge exchange in Swedish higher education policy. It is interesting for our purposes, since it presents an infrastructural argument and framing around communication and exchange.

An initial observation is that there is a slippage or transfer in the report from addressing infrastructure for humanities and social science collaboration and exchange to addressing (sustainable) infrastructure for the production and delivery of knowledge.[128] In this way, a model based on collaboration, communication, and exchange becomes an infrastructural model for the humanities more broadly, which is parallel to the public humanities becoming the humanities in many ways. Such slippage is problematic, as it seemingly indicates a large scope and a broad humanities mandate, while it, in fact, decenters other types of humanities work infrastructurally.

The report under review demonstrates considerable overlap with the framing of public humanities, the critique of which is relevant in this context. First, the strong separation between communication and scholarship—for example, evident in the recommendations given in the report—is problematic, as it risks reducing humanities infrastructure (and "exchange") to communication.[129] It is particularly unfortunate to separate communication from scholarship and teaching at a time when it seems that materials, research questions, and modes of articulation are becoming more integrated and codependent and when, to a larger degree, interpretation and scholarship are understood as embedded in the "product."[130]

Second, the collaboration described seems one-way and limited and is mostly from the humanities outward, essentially delivering knowledge and reaching out with knowledge produced by scholars that are set to analyze the world (a disengaged model).[131] This corresponds to the critique of public humanities as starting out from the academy rather than what is outside—what has been called a "missionary model."[132] Mary Mullen argues that public humanities scholarship needs to "allow community groups and public actions to transform how the university defines knowledge and culture."[133]

Third, this academy-centric notion of communication and exchange also construes the humanities as largely fixed, while the world is described as undergoing major structural changes (the media landscape, publication patterns and strategies, etc.). This view of the humanities as largely unchanged is directly reflected in the infrastructural aspirations. The static view of the humanities manifested in the report (and in much public humanities work) is often associated with conventional types of collaboration and a traditional model of the humanities. These are important, but the communication model of humanistic infrastructure often lacks a strong vision for what the humanities could be in the future. It is, for example, perfectly possible to think of humanistic collaboration in terms of a much more active and interventionist engagement in relation to societal and cultural challenges.

Interlude: Experiencing Centers and Institutes

There are humanistic infrastructure-like entities that are almost never called "research infrastructures" and will not normally be found on any infrastructure road maps. I will now consider one such institution in some detail. Humanities centers and institutes are platforms for supporting scholarship in the humanities outside the regular disciplinary and departmental structures, typically through offering fellowship programs, space, and different types of activities and programming. Some institutes of advanced study (IAS) also belong to this category, although they typically have a broader disciplinary scope. Such institutes and centers are particularly important to this book, as they sit outside regular disciplinary structures, have a connective ethos, and embody an infrastructural model relevant to the discussion of capacity building in chapter 4. I will first provide some personal reflections in this interlude, followed by an extended case study.

HUMlab at Umeå University, which I directed from 2000 to 2014, was rarely described as a center. The "lab" part of its name functioned as a default denominator, and when we presented the lab, we most often referred to it as a meeting place for culture, humanities, and information technology. But as we filled the lab with more content and structure, it acquired many of the properties frequently associated with humanities centers, including activities, programming, an international postdoctoral and artist fellowship program, space, and broad outreach.[134] Technology was an important part of the setup from the beginning, which set HUMlab apart from most humanities centers and presumably made it easier to eventually classify HUMlab as a research infrastructure. Curating and convening became important processes for implementing a crosscutting meeting place. Over time, a directionality developed around fostering and shaping dialogues and meetings, pushing questions at the intersection of criticality and technology broadly, and taking place at the table at the university, in the region as well as nationally and internationally.

This experience gave us firsthand experience of the challenges and tensions of being placed in between and being "center-like," including the work it takes to maintain that liminal position (organizational logic strives to normalize); how material implementations and conceptual grounding need to go together (in technological design and intellectual making); the necessity to have integrity as well as be helpful, humble, and curious (not being a service function); the difficulty in creating long-term career positions, even when there are resources (teaching and tenure mainly belong to the disciplines); and how institutional platforms have inflections regardless of their presumed openness.

There were successes and failures. Some failures include that while HUMlab was very open, it was a university institution that did not reach out equally across society. It was also more focused on epistemology than on societal engagement, which was not a failure but perhaps a missed opportunity. We created long-term positions for faculty in the lab, which was a good thing, but we were not really successful in finding ways of creating systematic career paths between the lab and departments. Among the successes, we built up an experimental meeting place, worked broadly with actors outside the university, and produced software as well as publications and a sense of belonging. We had a direction and became an internationally established place for thinking about the future of the field through making it happen. We also had integrity and an openness to different epistemological traditions, which helped develop a generative model of curatorship and constructive exchange.

As a young director, I traveled to see other environments, some placed in what we now call "digital humanities," but also other types of institutions and formations. This is how I encountered the network HASTAC (Humanities, Arts, Science and Technology Alliance and Collaboratory) and people interested in the kind of in-between work I found intriguing and exciting. One opportunity that came out of this was closer collaboration with different centers, including the University of California Humanities Research Institute (UCHRI), a UC-wide institution. One important aspect of UCHRI is that although based at UC Irvine (where I did a short-term "fellow" stay), it operates across ten universities and ideally leverages power across this system in a networked model. It also works with local humanities centers at the various universities. I felt that there was strong intellectual leadership and acuity at UCHRI that I had not felt at most of the digital humanities-style centers I had also visited.

In 2006, I was asked to participate in a summer institute under the rubric of "Cyberinfrastructure for Humanities, Arts and Social Sciences" at UC San Diego organized by UCHRI and the University of California San Diego Supercomputer Center (SDSC). The institute was supposed to:

> **introduce humanists, artists and social scientists to commonly used and emergent information technology tools and resources.**[135]

When I moderated the final conversation (which I was asked to do during the event), it became clear how the workshop in some ways had failed. Many of the humanists and social scientists were frustrated—even furious—and there was a sense among them that their research questions and concerns had not been taken seriously or even deemed relevant. The participants from SDSC had tried their best, but the setup was not conducive to rich or informed exchange.

In retrospect, I think the institute should have been framed and curated differently, including being clear about the learning process going two ways (less like a computer course meant to "introduce" humanists to technology and more like an exchange of different types of expertise and "content"). It would probably also have helped if "cyberinfrastructure" had been framed as a conceptual and sociocultural phenomenon, as well as a technologically driven system. Even so, the summer institute seemed worthwhile, and the final conversation aired out some of these tensions and felt like a step forward, albeit a hesitant step. It was an occasion to come together around emerging questions that needed to be facilitated, which is a crucial function of centers and institutes. This function—or perhaps better, responsibility—requires an ability to see where the action is or will be and where there are exciting, contentious, and productive "cuts," whether methodological, intellectual, or societal.

Much later, in 2015, I came to the Graduate Center at City University New York (CUNY) as a fellow at the Advanced Research Collaboratory (ARC). ARC had national or international fellows (like me) and fellows from the CUNY system, whose twenty-five campuses make it into the nation's largest urban public university. The Graduate Center, located on Fifth Avenue a couple of blocks away from the Empire State Building (and Macy's), is the privileged "flagship" campus, and conditions vary considerably across the system, not least in terms of teaching loads and resources. The different colleges are also embedded in different boroughs, which bring their own contexts and realities.

Being part of the CUNY system was central to the mission of ARC just like having a student connection. As fellows, we acted as advisors or mentors for graduate students (student fellows) who were active across the system. This was not a huge time commitment but felt meaningful and connective. Such connection making carries extra importance for a state school such as CUNY (mainly funded by tax money and tuition) and helps motivate the investment, particularly in external fellows.

ARC can perhaps best be described as a university-based institution of advanced study, but it is not clearly framed in that way, and it did not feel like an advanced institute. Whereas the Graduate Center building, a former department store, is grand, the premises for ARC were low key and unassuming. There was little there in terms of explicit or dedicated technology, which over time helped me arrive at a less technologically driven understanding of infrastructure. The center does not have an overarching theme but promotes certain areas that have resonance among CUNY faculty, including race, migration, and diversity and inequality. ARC also seemed much less isolationist than many other institutes and centers I had encountered. For example, it encouraged fellows to become part of NYC. Perhaps this was unavoidable, given its location in a vibrant metropolis, but it felt like a genuine sentiment. Although there was a weekly Tuesday tea and a seminar series, there was little control of fellows. In fact, the founding director, Donald Robotham, told me that they had chosen—despite some contradictory advice—not to "police" fellows.[136]

Each fellow had to give a seminar, and the one thing I remember best from my fellowship at CUNY is the first seminar in that series. Naomi Murakawa, who was a year-long fellow normally based at Princeton University, spoke about US police and so-called procedural justice under the title "The Perils of Police Reform." Procedural justice can be described as guides or instructions (processes) for handling challenging situations with empathy and through projecting understanding, striving toward racial repair through procedural legitimacy. Murakawa's critique of this approach was grounded in her earlier work on police and race. The windowless and hot seminar room was packed, and people were standing outside, in the lobby area, attending without seeing the speaker or her slides. There were many participants from outside CUNY and a vibrant rowdy energy. The discussion was intense, sharp, and, at times, heated, and it was much concerned with the world and making things better. To me, this was not just a seminar but an inspirational way to engage with the outside world.

In the spring of 2022, I was generously installed in a cell in a magnificent building—Linneanum—in the Botanical Gardens in Uppsala (figure 2.3). The construction of the building, which was started in 1787, was commissioned by King Gustavus III. I had become a fellow at the Swedish Collegium for Advanced Study (SCAS), a national center situated in Uppsala.

SCAS, instituted in 1985, is what Birgit Padberg calls a "second-generation IAS," which are independent and national.[137] The lineage goes back to the Princeton IAS (1930), which is a legendary institution ("one of the world's foremost centers for theoretical research and intellectual inquiry"), independent but without a formal national status.[138] In many ways, SCAS is the epitome of the traditional IAS: privileged grounds, fellows, offices, housing, compulsory lunches, a weekly or biweekly seminar, and proximity to a prestigious university (Uppsala University) but formally outside that university. There is a chef (who is fantastic), a palatial space upstairs (largely unused), guided tours to the countryside with the very competent university guide, and a special library delivery service from the university library (a couple of blocks away) to SCAS.

Being at SCAS was like being in a dream world, even if primarily designed for fellows working on their book manuscripts. Fellows would leave their offices at around half past twelve to walk down to the lunch premises and cluster around the round tables. There was a social program, tours, and a seminar series. Fellows were expected to be there, have lunch together, and attend the seminars. The social program was also semi-compulsory ("your participation is kindly expected"). I received support for a workshop I organized together with Arizona State University, which was much appreciated. There was a large degree of freedom and few obligations, but also a sense of being closed off, which was emphasized by the grounds, the locked buildings, and apartments for external fellows on or close to the main buildings.

While I felt incredibly privileged to be there, was energized by the international community of scholars, and appreciated the time given to me and

Figure 2.3
Office space at the Swedish Collegium for Advanced Study.

the warmth of the operation, I also came to reflect on why I did not feel I was in quite the right place. Perhaps my perspective is non-paradigmatic. SCAS is an excellent IAS with great staff and leadership. I think what I missed, though, was a soul associated with the grand infrastructure. The institute was by itself, in itself, and despite the historical surroundings, it could as well have been in another place, in another country; it was like a spaceship. There was little local or national connectivity. Also, because there was no clear intellectual direction beyond individual fellow projects and the general serendipity supported by the institute, the idea of the IAS became the primary anchor point, including the associated conditioning (e.g., individual projects). This idea is strongly connected to elitism and being selected—possibly even more so in a second-generation IAS—and perhaps I just did not feel I belonged there.

While at SCAS and afterward, I was excited about the potential for institutes such as SCAS to open up, have more directionality, using dynamic thematicization to allow nimble responses, reach outside, support an increased epistemological range, and be a unique platform for carrying out high-quality research as well as engaging intellectually and practically in an outward direction. These things are not mutually exclusive, and changes could be made without upending the basic model, but it may be that what I wanted and felt in relation to SCAS was still too much outside the institutional and epistemological tradition to be possible.

Case Study: Humanities Centers and Institutes

As humanities centers and institutes embed ideas, views on knowledge production, and an operation built around spaces, tools, and other resources, they are ideal for a critical case study of implemented humanistic infrastructure.[139] They also have the potential to be "integrative niches," just like the emerging humanities discussed in chapter 3, potentially providing building blocks for humane and humanistic capacity building as envisioned in chapter 4.[140] Although the focus is on humanities centers and institutes, I will also consider institutes of advanced study, as they bear some similarities to humanities centers.

History and Ontology

Humanities centers are most strongly established in the US, where they have a history that extends over more than sixty-five years. The first campus humanities center, the Institute for Research in the Humanities at University of Wisconsin at Madison, was founded in 1959.[141] It was followed by the Society for the Humanities (A. D. White Center) at Cornell and the Humanities Center at John Hopkins in 1966.[142] The introduction of humanities centers in the US came in three waves: an elitist stage from the 1960s, a broader public engagement in the 1980s, and a more general and administrative uptake since the 2000s.[143] An early example outside of the US is the Humanities Research Centre at the Australian National University, which was established in 1974 but detailed in reports as early as 1963 and 1969.[144] Most humanities center-like institutions in Europe, however, were established in the 2000s, including CRASSH, University of Cambridge (2001), and the Long Room Hub, Trinity College Dublin (2006).

The landscape of humanities centers and institutes shows considerable internal variation.[145] A US-based account from 1980 lists more than a hundred centers of different types, including the Center for the Study of Democratic Institutions, the Center for Southern Folklore, and the Japan Institute.[146] We can also think of digital humanities centers and cultural institutions outside of the university, such as the Huntington Library with its programming and fellowships. In this case study, however, I will mainly consider institutions part of the trajectory and typology indicated above.

Additionally, there is partial overlap with advanced institutes—some of which, in part, have a humanities orientation—that date back to the 1930s with IAS Princeton.[147] In contrast to early IAS institutions, as Britta Padberg convincingly shows, there has been a move toward institutes of advanced study that are university based.[148] These would seem to have more in common with humanities centers, although with a broader disciplinary focus. According to Padberg, in contrast to independent institutes of advanced studies, third-generation university-based institutes are more geared toward ideas and themes.[149] They are also often more outward-reaching and connective.

Conditioning and Framing

The underlying agenda paired with a particular infrastructural configuration creates opportunities and conditions for scholarly work:

> **The Humanities Center is a little corner of paradise for giving oneself fully to one's project.[150]**

This statement from a fellow at the Stanford Humanities Center is representative of the laudatory narratives often surrounding centers and institutes. Fellows, who are given precious time and support, often become part of the narrative of the center. What is there not to like?

However, we are clearly concerned with opportunities for select groups of people. Humanities centers and institutes are generally privileged institutions that offer opportunities for highly ranked visitors and members of their own university or university systems. Some are distinctly elite institutions. The Whitney Humanities Center at Yale University was once described as "a gathering place for overlapping excellences."[151] The IAS Princeton welcomed a new cohort of fellows in 2022 through identifying them as "250 of the planet's most interesting scholars."[152] Wealthier universities and university systems are more likely to have well-resourced centers than others. Adjuncts are less likely than professors to get (or be eligible for) fellowships.

The conditions for knowledge production offered at humanities centers privilege certain types of projects, participation, scholarly processes, and modes of engagement. For example, individual writing is normally expected rather than collaborative interventionistic projects. The book is the gold standard for scholarly output. The conditioning—in terms of both opportunities and constraints—permeates the whole configuration of the center.

Conditioning is negotiated to some degree. For example, the establishment of themes and programs cannot easily be done from the top down without negotiation, especially as most centers depend on local faculty. A previous director of the Wesleyan Humanities Center, Hayden White, said at a 1976 meeting that the center had been unsuccessful in developing programs concerning popular culture because the humanities departments had a primary commitment to the canon of standard works and limited interest in critically assessing the effectiveness of teaching.[153] The intermediate and curatorial role can be challenging and quite time consuming. The first director of the Humanities Research Centre at the Australian National University, Ian Donaldson, expressed some frustration with participation from the arts:

> **At times—to put matters bluntly—the Faculty doesn't seem very interested in what the Centre does: in the visitors it brings to Canberra, and the conferences and seminars it runs; doesn't respond to requests for suggestions concerning possible future visitors and conferences; *doesn't much want to be stimulated.***[154]

1442	45	1673	227
Fellows in residence	Fields represented	Books resulting from fellowships	Fellows from international institutions

Figure 2.4
Statistics from the National Humanities Center.

As noted, humanities centers and institutes tend to have a strong focus on monographs and books. The basic idea is that the humanities center offers a space—a refuge in terms of time and place—needed for scholars to finish (or start) significant monographic projects. The number of monographs and other books "produced" during fellowships provides a key metric for many centers and institutes.

The Stanford Humanities Center, for example, only lists book publications in their annual reports.[155] The physical materiality of the publications is emphasized through ample photos of books in the reports, often with artificial depth added to mimic a printed book (rather than showing a flat cover). In a similar vein, the Institute for Research in the Humanities at UW–Madison features a photo of a bookshelf with books on the cover page of their website as well as a top-level menu category for books.[156] The number of books over time matter, and the National Humanities Center lists 1,673 books as a key cumulative result as of February 13, 2021 (figure 2.4).[157]

For many centers, monographic production provides the key metric for success, together with the number of fellows hosted and events organized. The disciplinary and institutional range—the number of different disciplines and institutions—is another source of metrics.

It is rare to find humanities centers that provide extensive infrastructural support for advanced digital production, critical making practices, or lab-based collaborative work. This pattern can also be seen in centers not exclusively focused on the humanities. Owen Meredith Wilson, director of the Stanford Center for Advanced Study in the Behavioral Sciences from 1966 to 1975, is reported to have said about their fellow selection:

> **The white-rat man must be at the point where he can leave his rats home.**[158]

Similarly, Swedish SCAS states in the description of its natural sciences program that there are no research laboratories at the collegium, and although there are facilities in the vicinity, "it is envisaged that scholars while in residence will primarily use such facilities as resources for intellectual encounters and dialogues rather than to conduct their own research."[159] A conditioning of knowledge production is occurring through this cultural subtext, whether enforced or not.

This separation of analyzing and writing from creative, collaborative, and making practices is relevant to the development of centers and institutes at a time when there is increased need for humanities-driven making practices in digital knowledge work, mixed-media expressions, and interventionistic approaches beyond what is labeled public humanities initiatives and for responding collaboratively to societal challenges and complex problems. This separation has a long history in the humanities, as exemplified by the separation of art history and artistic practice.[160] Similarly, centers support certain modes of engagement, scholarly processes, and results better than others through the available infrastructure.

Institutional Position and Governance

Institutionally, humanities centers are typically placed outside or between conventional disciplines and departments, which, among other things, allows them to operate across disciplines and themes. This intermediate placement is often seen as central to the operation, and managing such a position is directly relevant to the consideration of arenas for transversal knowledge production in chapter 4.

In 1988, at a keystone meeting at UC Irvine about humanities centers, one long-term director emphasized that a center "must categorically exist independently of departments."[161] They are rarely independent, quite the contrary, but the division of roles is important. For example, research produced through centers does not accumulate there as much as in the home departments of the faculty engaged in the operation (whether through a fellowship, attending a seminar or being part of a public humanities effort). There is an accumulation, however, of in-depth competence about exchange around intellectual issues, facilitating interdisciplinary processes, and coordinating grant proposals and managing large programs. Such factors will be discussed under the rubrics of curatorship and arenas for transversal knowledge production in chapter 4.

Another director at the conference in Irvine pointed to how it may be mutually beneficial to have a direct connection between a center and university-level leadership. In this way, rather than competing primarily with humanities departments, a humanities center can potentially access other funding streams and modes of institutional recognition, which, in turn, can give a certain degree of freedom. And university leadership can get both "a piece of the intellectual action" and larger long-term control of resources.[162] This high-level interest, the outside position, and the thematic breadth often contribute to making humanities centers strategic and iconic rather than mundane and conventionally institutional.

Unlike departments, whose governance normally includes faculty, centers often have a more singular and direct leadership, which can be very long lasting. Several centers in the US have had the same leadership in the range of twenty years.[163] As Eugene Goodheart put it at the UCI conference, institute leaders should have an "instinct for where the action is."[164] They can help anchor an institution intellectually beyond the annual cycle of fellows and events and, together with their colleagues, shape the institution's texture and direction, both through the practice of the center and through their own scholarly work. In many cases, institute leaders also get to represent the humanities and are invited to outline intellectual and institutional paths forward in a world where many other institutional representatives (deans, research councils, etc.) often have to be more constrained in their future projections for political, collegial, and strategic reasons.

Space and Time

The iconic quality of centers and institutes can be emphasized by the space of the operation. As noted, many centers have their own space, which can be used for meetings and events and as workspace for resident scholars (and to dazzle visitors), as evident in these photos presented on the website of the Center for the Humanities at Wesleyan University (figure 2.5).[165] These spaces are often intimately connected to the identity and operation of the center, especially for "prestigious" centers (not only the older ones).

The connection between space and identity is sometimes made explicit, as when the Franke Institute describes itself as "both an idea and a place."[166] Kathleen Woodward emphasizes how humanities institutes have

Figure 2.5
Center for the Humanities at Wesleyan University. Photo credit: Wesleyan University.

an intermediate function within the institution and can create a "unique sense of place."[167]

These institutions extend not only spatially, naturally, but also over time. Infrastructure is temporal as well as spatial.[168] Infrastructures are expected to be long-standing, and the investments are motivated by the longevity of whatever is being built. A humanities center or institute is usually meant to last for quite some time. It is not a project, although it is likely not taken to be as stable as a department.

A key temporal quality of many humanities centers is that they give substantial time for scholars to work on their projects, whether through fellowships or other means. They offer space, resources, and an intellectual milieu for a certain amount of time. The time factor—being able to focus on scholarly work—is invaluable to many researchers. The time given is also protected by the center or institute. Humanities centers remove the humanists from their local context and the many associated commitments. This removal does not result in a full disconnect from institutional realities, however, and the product envisioned (usually a book) is embedded in a disciplinary and departmental context and is needed for tenure, accreditation, or recognition.

Conventional humanities centers also introduce an alternative or additional type of temporality to institutional life through themes, programming, and fellows. Themes introduce intellectual threads that may be active for a semester or a year, programming can create times of intellectual coming together and is often done with a certain regularity, and fellows—internal and external—change the rhythm and concentration of the institution over a set period of time. As Juliet MacCannell emphasizes, humanities centers can break up the temporal and organizational stability of departments and disciplines.[169] This can also be a competitive situation if time and attention are seen as divided between departments and centers.

Centers and Institutes as Infrastructure

Humanities centers make research possible and, to some degree, other aspects of academic work and public engagement in a robust manner beyond the scale of individual researchers, themes, and project timelines. Just like many science-based infrastructures, they are often described in terms of being creative, interdisciplinary, and dynamic. In fact, in most ways, they would seem to be a good fit for scientifically driven notions of research infrastructure, such as this definition by the European Union: "facilities, resources and services used by the science community to conduct research and foster innovation."[170]

A key difference, however, is that centers are normally not presented or packaged in terms of technology or specific instruments. As I have argued elsewhere, there are no obvious humanistiscopes.[171] The European Spallation Source, which will be a diverse human, programmatic, and technological facility, is described as "the world's most powerful neutron source" and is compared to a large microscope.[172] This reification of complex operations is central to infrastructure work (chapter 1). Additionally, an instrumental view of research infrastructure tends to go together with a focused purpose and problem definition. The ESS website tells us that the facility will make it possible to study different materials in order to "see and understand basic atomic structures and forces."[173] In contrast, humanities centers are often not focused on specific topics, questions, or disciplines—at least, not over time—and do not normally set out to offer direct solutions to scientific and societal challenges.

There is a strong case to be made for humanities centers and institutes of advanced study as infrastructures. This argument is analytical, as centers fulfil many infrastructural criteria. It is also a policy argument in the sense that centers or networks of centers could well be part of research infrastructure road maps and infrastructure policy work. There is a constructive argument too, which is particularly important to this book and which necessarily builds on the analytical perspective.

The deconstruction of infrastructures can help build new ones. Humanities centers and institutes are recognizable platforms with a range of useful properties and extensive experience. While we may not be primarily interested in privileging traditional modes of scholarly knowledge production or creating a local "haven," we can draw on their generic and integrative qualities as well as a well-developed range of tools, processes, and techniques, all of which can contribute to building humane and humanistic capacity—the topic of chapter 4. In chapter 3, we will consider where such tools may come from and how they can be used for the active humanism needed for such capacity building.

3

Instrumentalizing the Humanistic

This chapter considers how humanistic and humane methods, perspectives, and ideas can become part of a repertoire or a toolset used to respond to academic and societal challenges as part of an active humanism. Such a toolset should be built around approaches where the humanistic element suggests an opportunity to change and shift the world in meaningful ways. While a toolset obviously is a set of tools or approaches, it is as much a set of mediating strategies and processes for supporting and facilitating collaborative critical-constructive work.

The primary interest of this chapter is not single methods or perspectives, such as terraforming, designing for difference, staying with the trouble, or infrastructural inversion, but rather how we can build capacity across different approaches.[1] How can these and other approaches inform each other, and how can they be instrumentalized as part of humane capacity building for responding collaboratively to academic and societal challenges? One central assumption is that building such capacity requires working across domains, facilitating respectful dialogue that can inform action, using the humane and humanistic as a driving logic, and being willing to change, critique, and build together.

While the language and ideas of instrumentalizing, aggregating, building capacity, and responding to societal challenges may be counter to the grain of the humanities in some ways, the question posed above is not meant to provoke. Rather, it is profoundly and necessarily driven by the humanistic and humane in at least three ways.

First, we need critical knowledge to more directly inform the ways in which we imagine and build a more humane world. Knowledge production is a collaborative process across modes of engagement that feeds back both into the world and scholarship. Second, there is a growing number of people, not least early- and mid-career scholars and experts across many humanistic and related fields, who are looking for ways to connect their critical work more systematically to constructive action (and vice versa). Many are also looking for places, models, and support for doing such work effectively and sustainably. Third, instrumentalizing in the sense imagined here is as much a way of encouraging and challenging humanistic scholarship and scholars as it is to package and present the potential of the humanistic and humane.

What tools might there be? Two examples discussed in the following are terraforming and infrastructural inversion (or turning things upside down). But for such and other tools to find a place, the humanities need to have a repertoire that extends beyond studying, interpreting, and critiquing. Some fields in particular show potential for extending that repertoire in a more active direction. These include emergent humanities—for example, environmental humanities and urban humanities—as well as fields such as disability studies and queer studies. The fields, just as conventional humanities disciplines and domains outside the humanities, are needed for the capacity building envisioned in chapter 4. As we will see, however, considering the instrumentalization of humanistic and humane knowledge is a question not just of institutional or disciplinary engagement but also of individual motivation. Building toolboxes requires both, as well as a critical, creative, and constructive ethos.

Planetary Perspectives

Literary scholar Ursula Heise considers how "terraforming" as an analytical lens and imaginary tool can challenge assumptions about the city as an anthesis to nature and the Anthropocene as human centered, potentially opening up for a step toward "a more-than-human democracy."[2] Terraforming becomes both a literary technique and a way to consider urban ecology and agency, as in Kim Stanley Robinson's science fiction novel *2312*:

> **As a complex ecological system, then, the city in Robinson's novel provides a template for thinking about terraformed planets and the transformation of nature through human agency.**[3]

Terraforming can thus function as a tool to reconsider human-nature interfaces in urban landscapes. Under Heise's analysis, terraforming becomes a means to rethink the Anthropocene as a concept and some of the core assumptions of environmentalism in relation to our own planet. Anthropologist Joseph Masco also employs terraforming as a conceptual tool—although grounded in the history of nuclear engineering (which was arguably pitched as a terraforming strategy)—to connect the rise of a planetary imagery and optics to an origin in the Cold War.[4] He devises an age of fallout, of unintended consequences. Planet Earth has been subject to a long-term terraforming project, driven by industrialism, capitalism, and militarism:

> **Constituted as a side effect of the industrial age, it [climate change] articulates the ongoing challenge to conceptualizing an ecological security that is not based on the split between the engineered event**

and its unwanted aftermath, between the boom and the bust of capital, between the pre-emptions and blowbacks of counterterror.[5]

Although Masco's article is mainly critical and historical, it clearly has a constructive ethos, as can be seen in the daunting challenge presented toward the end of the article: the need to conceptualize a viable model of ecological security. The same is true of Heise's interest in reconsidering (and possibly reshaping) urbanity. It is nevertheless a stretch, perhaps, to think that terraforming would be a conceptual tool used by governments, organizations, and corporations when approaching environmental, ecological, and urban challenges. However, as I will argue, terraforming could be one tool among many driven by the humane.

However constructive, this suggested engagement is rather different from the much more design-oriented framework suggested by sociologist and visual arts scholar Benjamin Bratton in the project "The Terraforming." Here, the goal is to "fix" the Anthropocene ("a headless terraforming gone wrong") or, more precisely, to "fundamentally transform Earth's cities, technologies, and ecosystems to ensure that the planet will be capable of supporting Earth-like life."[6] According to Bratton, the solution has to be anthropogenic, interventionist, and planned, driven by artificiality, astronomy, and automation. There is hope here, partly fueled by the technological sublime, as well as directionality and conceptual situatedness.[7]

Nevertheless, "The Terraforming" project would not seem to lend itself to being a tool for capacity building as discussed in this book. The framework already seems fixed, not readily open to questioning, and it may be too much on the constructive, technological, and solutionist side of things to incorporate certain types of criticality and humaneness (including more-than-human humaneness). And while it is fruitful to consider the concept of "climate change" as an "epistemological accomplishment of planetary-scale computation," the interest does not primarily seem to be to alter or change the way that computation is carried out to allow for other or altered conceptual models and ways of "working" climate change.[8] There is a "big project" and "technological sublime" sensibility here that can be related to other big projects and infrastructures, including skyscrapers, dams, electricity, "smart cities," and the cyberinfrastructure movement referenced in earlier chapters, which was related to a previous large modernist computing project: the supercomputing movement.[9]

As Bratton's work demonstrates, planet-level engagement is becoming a more urgent issue, including geoengineering initiatives such as solar engineering and putting aerosols into the atmosphere to reflect sunlight and reduce global warming.[10] There is massive critique against these kinds of interventions, and the human, ethical, and social repercussions of such potential projects are immense, as is the potential. It seems clear that humanists and social scientists must be involved in such endeavors one way or another.[11]

The same is true of issues such as planetary management, planetary health, and planetary challenges, which are starting to appear as academic and societal endeavors, for example at Arizona State University and Stanford University.[12] Such enterprises will need to embed the kind of criticality that Heise's and Mosco's work manifests—for example, addressing the potentially problematic human-centeredness and arrogance of many such approaches as well as ethical implications. Critical knowledge also needs to be part of imagining "the future of the planet" (Stanford University) and figuring out what it might mean for a university to be like "a medical center for the planet" (Arizona State University).[13] The humanities must be potent enough to both critique and constructively engage with such notions and frames.

Turning Things Inside Out

In a chapter on energy humanities as a field, we are told that the point of energy humanities is to turn phenomena such as environmental degradation and global warming "inside out."[14] Another observer states that an important role for the humanities is to make implicit assumptions explicit—for example, questioning unrestricted economic growth as a given in European funding programs for grand challenges.[15] Turning things inside out seems to be one of the things the humanities do through their deconstructive and critical practices. Such work can potentially open a space for new interpretations and help us shift how we think about the world.[16] Turning things inside out can be regarded as a tool, even if a very general one. An important question then becomes how (and if) it connects to action and change.

Infrastructural Inversion

Geoffrey Bowker's notion of infrastructural inversion is a specific (if also fairly general) framework for turning things inside out that has been widely influential. In later work, Bowker and Susan Leigh Star defined infrastructural inversion as "recognizing the depths of interdependence of technical networks and standards, on the one hand, and the real work of politics and knowledge production on the other."[17] Infrastructural inversion was central to their seminal work on categorization and standards through instrumentalizing and articulating "a double vision," which was necessary to see the technologies, standards, and arrangements that were part of infrastructure but tended to disappear from view.[18]

In his analysis of the Schlumberger method for testing potential oil fields, Bowker introduced infrastructural inversion as a means of putting the organizational and cultural aspects at the forefront as opposed to only pure science or the technology.[19] He shows how the success of Schlumberger as a company came down to their ability to bundle technical and organizational features, developing a rhetoric, modifying the measurements, and changing the local environment of the oil well as well as the organization of oil companies. Yet, on the surface, the Schlumberger story was one of producing scientific measurements through pure scientific discovery.[20]

Paul N. Edwards argues that we need to invert an infrastructure to understand it, and in his important study of environmental data, he claims that data do not become data until you have turned the infrastructure upside down to see how it works.[21] This is done not only by the scholar studying the infrastructure but also by scientists and others involved in the operation. For example, people collecting environmental data over long stretches of time comment on how the data are disparate as well as dependent on local conditions and changing measurement standards.[22]

Infrastructural inversion as an analytical tool can—and often does—change our understanding of that infrastructure in significant ways, not least through channeling a mass of critical knowledge about infrastructures built up over decades. Revealing what lies underneath or explaining how something works (or not) is not the same thing as changing it, however, and infrastructural inversion does not typically feed into the active reimagining, reconfiguring, or building of infrastructure.

There are several reasons why this may be the case. For one thing, we might not expect infrastructure builders to read scholarly articles, especially not humanities and social sciences scholarship. Also, critical work of this type tends to stay in a deconstructive modality, which means that it is not obvious how it can inform making. The step is made larger because critical scholarship on infrastructures tends to be removed from the scholar, whether scientific or civic infrastructures. In such situations, the scholar normally has an outside position and little (assumed) control over the infrastructure.

I argue that humanistic tools and methods of this sort can be used actively to change infrastructure. The point is not that all critical work can be used to inform and enact change but rather that there is an opportunity and need to move in a constructive direction at some scale. One starting point for such an epistemological and ontological reorientation can be approaches such as infrastructural inversion, especially if they become part of a capacity designed to extend such critical work.

Looking Inward

A lived and situated perspective can arguably make it easier to move toward constructive modes of engagement. Seeing infrastructural inversion in your own local context can help negotiate complicity and potentially engage change there as well as in the broader societal context.[23] This is an argument made by research scientist Rida Qadri (following Noopur Raval):

> **Critical tech scholars must turn the critical lens inwards to investigate our own research practices, methods and the power asymmetries we perpetuate.**[24]

In his discussion of this inward perspective drawing on Donna Haraway's work on seeing and "bloody eyes," cultural anthropologist Joseph Dumit postulates a series of interventions under the rubric "the implosion project," an educational approach in several steps suggestive of infrastructural inversion as a tool.[25] At the center of the approach is a specific object, process, artifact, or fact that students get to explore. Among the dimensions Dumit explicates (each with a series of questions) in the initial step are labor, professional/epistemological, material, technological, bodily/organic, political, economic, historic, mythic, and symbolic. Students are asked to produce a knowledge map around these dimensions. A series of steps takes the analysis further. One example is that the student participants are expected to consider what they do not know and to look at possible explanations for patterns of ignorance (e.g., difficulty, unimportance, inefficiency, or exhaustion). We are reminded that infrastructural inversion (and, more broadly, deconstruction and institutional critique) does not necessarily apply as easily or equally to all parts of the infrastructure under scrutiny.

The implosion framework is powerful because it is theoretically anchored, detailed, and forces engagement with one's own infrastructures and complicity. It should be noted, however, that despite the fact that the framework is "active," it essentially comprises an analytical deconstructive approach. Nevertheless, it is easy to see how it could be turned into a constructive approach as a tool by itself or under the rubric infrastructural inversion, especially given the conceptual and material detail provided by the framework.

Infrastructural Inversion as a Constructive Move

While infrastructural inversion has been primarily construed as an analytical tool, it clearly has constructive potential in relation to both academic and civic infrastructure. Being able to systematically take apart infrastructural systems can be useful for considering how to develop, reimagine, and build infrastructures. Bowker's work on Schlumberger, for example, could be used to inform the humanities or urban initiatives about how to shape their infrastructure and infrastructural policies. What is the bundling of features needed? What is the rhetoric, story, and science of humanistic infrastructure or of a humane city?

Similarly, the understanding of environmental data that Edwards's work provides through infrastructural inversion could be employed to rethink data models and narratives needed to approach climate change as a planetary challenge. For example, how do we aggregate qualitative data that are local and situated in meaningful and actionable ways? How can consolidation of data be made more transparent? Scholarship on topics such as the history of scalar mediation and optimizing of urban data could feed into considering how we build alternative models.[26] Again, moving in such a direction requires us to take a position in relation to the current regime of data: not only to point to the complexity of how data get produced but also to suggest how different notions of data (potentially also including narrative and experience-based materials) and data enactment (what we do with data, how we inform action) can help us approach climate change constructively. The insights produced by Edwards and other critical data researchers can directly help us in this endeavor.[27]

Infrastructural inversion or turning things inside out can be applied not only to infrastructures constructively but also to infrastructural frameworks or infrastructure as a system. If infrastructure is normatively taken to be big, technological, standardized, homogenous, and long term, these basic conditions can or perhaps should be challenged. Infrastructure as a framework could be shifted to include the humane as a driving force, and infrastructure could be imagined (if we so like) as small as well as big, transient, not as standardized and less homogenous. Again, imagining alternative conceptions of infrastructure is a constructive move, even if the frameworks are not shifted, but the point is also that change is possible and is made possible through critical and imaginative work.

The Repertoire of the Humanities

Taking the step from critical deconstructive scholarship to actively redesigning the object of study is challenging for several reasons both internal and external to the humanities. Humanists are normally not taught or expected to engage in the building of infrastructure, such as sociotechnological systems. The field, admittedly heterogenous, is typically presented as analytic rather than actively constructive. This positioning is reflected in the framing of the humanities, which is much more likely to include "study" or "write" than "intervene" or "build." These verbs can be said to invoke different modes of engagement—essentially ways of doing work and engaging with the world—and, in some cases, different epistemic traditions. I use the term "repertoire" to describe the epistemic range of an area, in this case the humanities.

The material used in the analysis below consists mainly of descriptions of the humanities, reports, calls for funding, and presentations of individual projects. We need to be aware, however, that such descriptions have different functions and that we are concerned with discourse rather than actual humanities work. Nevertheless, institutional statements are not just ephemera or PR but rather represent and manifest the institutions in question. It is also important to acknowledge the vital work the humanities are doing as public intellectuals, critics, and collaborators. The point is not to suggest that constructive engagement is better than analytical engagement or vice versa but rather to nuance the analysis and suggest that there is room for the humanities to extend their repertoire.

In a landmark report, "Humanities World Report 2015," the authors draw on a literature review, interviews, and workshops to identify a set of answers to the question of the value of humanities research and what they call "a plausible taxonomy of the most prominent attempts to articulate the value of the humanities around the world."[28] Among the categories listed are "intrinsic value," "cultural heritage," "economic value," "contribution to other disciplines," "critical thinking," and "aesthetic appreciation."[29] Each category is accompanied by a brief description (as well as a more extensive account later in the report), and among the verbs—in their basic form—used to describe the role and value of the humanities are "have a value," "benefit society," "help create tolerance," "aid decision-making," "benefit society through challenging established positions," "feed into other fields," "promote a culture of innovation," and "enhance one's personal and spiritual wellbeing."[30] There are clearly degrees of agency and activity in the descriptions, which could be used to suggest a simple scale of verbal agency ranging from "have," "benefit," and "enhance" to more active forms such as "aid," "promote," and "challenge." In general, the humanities are given a distanced and indirect role, even when more engaged, as in "help create tolerance," and relatedly, they are assigned an auxiliary or supplementary role when said to "aid decision-making" or "feed into other fields." It is not that the humanities as presented in this material do not have power and ability but rather that the repertoire still seems relatively limited and distanced.

In an analysis of data from web-based presentations of the humanities from thirteen Swedish comprehensive or technological universities, the most common verbs used to describe what the humanities do and offer are "study," "analyze," "explore/investigate," and "understand."[31] Others include "problematize," "organize/systematize/process," "identify/find problems/answers," and a cluster of verbs to with explaining and presenting: "explain," "give/have insights/understanding," "create understanding," and "give an overview." In general, the presentations, many of which are intended for prospective students, represent a conventional view on the humanities although with an increased weight given to the societal and professional potential of the disciplines. There is considerable variation between different universities in the material—for example, ranging from elaborate frameworks for explicating humanistic competence to very simple and brief statements.

This emphasis on relevance in higher education can be traced back to the 1970s or 1980s in many industrial countries and is linked to a prolonged debate about the role and function of the humanities.[32] In Sweden, innovation has been a central dogma for at least two decades in higher education policy.[33] Three of the universities in the study state that the humanities analyze/understand and solve problems (or that the programs produce knowledge or methods used to solve problems). There are also two cases of "realize" in relation to ideas and sustainable societies. "Solve" is the most agentive verb in these data and seems almost like an imposed formula without much precision, perhaps akin to the kind of report language Kathleen Woodward calls dull and listless.[34] In contrast, the vocabulary for understanding, analyzing, and studying is much more developed and has more precision.

Progressive but Still a Limited Repertoire

We will look at other types of framings of the humanities later in this chapter, but the picture given above probably holds true for much of the conventional humanities. This repertoire is also used by more outward-leaning and interdisciplinary initiatives such as the Center for the Humanities at the University of Wisconsin–Madison:

> **The Center for the Humanities promotes the cross-disciplinary, collaborative, and public humanities across and beyond the University of Wisconsin–Madison. We support traditional and new forms of inquiry in the humanities and engage the public through partnerships that broaden the ways that knowledge circulates.**[35]

The center obviously does important work and not only seeks to support new forms of inquiry but also maintains the positionality and repertoire of the established humanities through talking about "inquiry" and broadening "the ways that knowledge circulates." It is also symptomatic that it is the humanities that engage the public rather than the opposite.

WASP-HS—a massive (and laudable) Swedish funding program for humanities and social science research on AI, autonomous systems, and software—provides a different example with a similar framing. According to a constitutional document at the host institution, the overall rationale is to support such research so that individuals, society, and industry can meet challenges and consequences of AI and autonomous systems in the best possible way.[36] This is somewhat similar to the use of "solve" in the humanities presentations discussed above. Further down in the document, the program is described as an addition to the technological research already supported (through WASP, which is a much larger program). The technologies in question (AI and autonomous systems) are said to have significant consequences for our society. WASP-HS is expected to explore ethical, financial, labor-related, social, and legal consequences and aspects of AI and autonomous systems and provide insights as to how future technological changes can be managed.

The humanities are being added to an existing program, which is a good thing, but the structure also reflects a technological imperative. The technological changes come before the humanities, and investments in technological research predate investments in research based in the humanities and social sciences.[37] The humanities are there to help control technology and to provide illumination. Despite the ambition to inform technological development, there remains a gap in the description between this desire and the rigor of more detailed framing. Constructions such as "explore" and "provide insights" seem somewhat weak and epistemologically conventional. The language used leaves unclear exactly how the connection between this epistemic culture and tradition and those of the technological work will come together productively.

This situation is not unique to WASP-HS. For example, the EVOKE Lab and Studio at UCI engages with social issues and concerns through creating "technologies and artifacts that evoke thought, debate, discussion, and consideration of a more harmonious mediated world."[38] Similarly, the Collaborative Future Making Network at Malmö University was described this way in 2018:

> **The Collaborative Future Making network (previously known as Alternative Future Making) addresses the crisis of democracy in the face of the multiple crises of our age. It will do so through refining participatory processes in which alternative futures can be articulated and debated.[39]**

Even these initiatives—which are more explicitly about direct change than WASP-HS—tend to stay in the discursive, imaginative, and prototyping frame. However, as Shaowen Bardzell and Jeffrey Bardzell importantly remind us, there is considerable value to the "effort after meaning itself": the marshaling of resources, the grappling, and the integration of such efforts in other processes.[40] This does not mean that there is not potential to push work at the intersection of critical and constructive engagement, exemplified by these valuable initiatives, further into a more operative and direct frame.

The question is not whether the humanities should analyze, explore, and contribute to our understanding of the human condition—this is what the humanities do—but whether the repertoire and frame of the humanities cannot be meaningfully extended. What about imagining, designing, building, making, intervening, disrupting, or changing? The way we present the humanities comes with responsibility and is a matter not just of discourse but also of the way we imagine and enact the humanities and humanistic work.

Locating Potential Capacity

There is no easy answer to the question of where to find the humanistic and humane knowledge needed to build capacity collaboratively to respond to great challenges and complex problems. It does not fully reside in the humanities as conventionally organized into departments and disciplines, although these are central to any such effort. It does not sit in any applied center oriented toward the humanistic, although processes and methods from such centers can be vital to tackle challenges. It does not entirely exist in engineering and science departments, although these can be important for moving research in a constructive direction at scale and for shifting educational paradigms. It cannot be found solely in NGOs and activist networks, although these know how to transform thoughts and ideas driven by the humane into action. It cannot be identified fully in civic organizations and institutions, although they need to be part of change and bring a unique sense of responsibility to the table. Also, it does not sit in corporative environments, although they can be key stakeholders and implementors of change. Furthermore, it lives in people as much as in institutions.

Building capacity for humanistic and humane knowledge should be open to all types of relevant work, for example disciplinary humanistic inquiry (including "singular" scholarship), design-oriented practice, and work carried out by emergent fields such as energy humanities and by outside fields and organizations. This openness is central to the transversal arenas and capacity building that will be discussed in chapter 4. There are some areas that are more likely to be starting points, however, especially for the instrumental perspective at heart of the current chapter.

These areas encompass emergent humanities—including urban, digital, environmental, and medical/health humanities—and fields such as disability studies, queer studies, and food studies. Heise identifies a series of fields that add either "humanities" or "studies" to a "concept that has in the past been the purview of disciplines outside the humanities and qualitative social sciences."[41] While these fields have their own complex histories and trajectories and cannot be regarded as homogenous, they generally share an interest in infrastructures and the lived world. This interest means that we are likely to find the kind of material, social, and organizational engagement and experience there that can be helpful in considering how to shift those infrastructures and conditions.

The Emerging Humanities

Pressure from Outside and Inside

The humanities are often taken to be static—indeed, they are expected to be unchanging and traditional. At the same time, the humanities are being put under considerable pressure inside and outside higher education. This pressure can be of many different types. The humanities have limited political leverage compared to many other areas. They are not seen as useful, innovative, or generative. Humanistic education is not believed to lead to (useful) employment in the same way as other fields, and they are associated with issues that are seen as problematic by some (such as identity politics). For example, the humanities are rarely included in national innovation strategies or seen as a driving force in research bills, and they are sometimes portrayed as dangerous, as in the discussion of critical race theory in the US.[42] Another source of pressure comes from the sense of the humanities as a unique enterprise under threat—in crisis—which can lead to a defensive and isolationist stance.

All these points can be contested or challenged, but there is no way of getting away from the sense of pressure, even in a country such as Sweden where the humanities (just as higher education generally) are in a relatively good position.[43] The usefulness of humanists has long been questioned. In Sweden, for example, strategies for restricting the number of students in the humanities have been discussed repeatedly through history.[44] As such examples demonstrate, the overall frame (the peripheral position and relative uselessness of the humanities) and supportive arguments (a pastime occupation for the wealthy, not leading to jobs, absurd identity politics, boutique research projects, etc.) have had strong currency regardless of their validity. Again, it is important to stress that all national and local contexts are different and that not all "humanities crises" are the same.

There is also a sense within the humanities, at least by some humanists, that how the humanities have been institutionalized is in conflict with contemporary needs and their own research and education interests. The infrastructure of the humanities—essentially understood as the solitary scholar model (chapter 2)—is seen as an obstacle to carrying out humanistic work, especially when such work is interdisciplinary, collaborative, and experimental.[45] The sharp separation of the humanistic disciplines—the assertion that these have their own domains and study objects—is also seen as problematic, and their study objects and methodologies are taken as not matching the world adequately anymore.[46]

This perceived misfit extends over several registers, including an inability to take on themes and challenges such as the city or the environment, a failure to engage seriously beyond the "human" as codified in the humanistic tradition, a critique of how things are being studied (e.g., close reading instead of quantitative methods, individually instead of collaboratively), and a concern with the limited scope as to what is being studied (e.g., conventional texts as opposed to multimodal representations).

It is not that the humanities have not supported work outside the realm of the disciplines, for example through thematic programs and through institutions such as humanities centers, but rather that these only make up a small footprint compared to the whole of the humanities (chapter 2). Furthermore, the more institutional of these formations may seem too conventional and restricted in terms of their epistemic, experimental, and infrastructural range, and less entrenched formations may have limited maneuverability and access to resources and be dependent on a department or discipline.

Fitting the Humanities

There is a range of humanities initiatives that resemble thematic centers and programs in some ways but also seem to operate outside this frame. They add "humanities" to some theme or perspective, such as "digital humanities" and "environmental humanities." Many of these varieties recognize the misfit sketched above and see a need to change the model to accommodate the themes and challenges pursued. The solution proposed is typically not changing the whole of the established humanities but rather building new platforms for humanities-driven work, whether inside or outside the institutional humanities.

In a book on urban humanities, we are told that the siloed configuration of humanities departments does not make any sense for approaching big urban questions.[47] The institution is simply not capable—as traditionally configured—to approach the topics and challenges at hand. Rather than adapting to these needs, the conventional humanities are seen as engaging with ever more specialized problems on a largely individual basis. Urban humanities, on the other hand, are said to engage with hyper objects such as "the city" collaboratively, experimentally, and speculatively.[48] Such modes of engagement, whether aspirational or implemented, are relevant for considering how to build humanistic and humane capacity.

One concern with conventional humanities as articulated in this and other descriptions is that the humanities are buried in their own disciplinary context and deconstructive frameworks and that they do not engage with the future constructively:

> **In engaging with future prospects, the urban humanities have introduced a way of thinking that stands some chance of breaking free from the hermeneutics of suspicion.[49]**

There is value in a hermeneutics of suspicion, in the critical capacity and propensity of the humanities, but the challenge is to recruit that capacity to imagine alternative futures as well as to be actively involved in the construction of those futures (also beyond "thinking"). What fields such as urban humanities offer is a sense of future making and a hope to potentially make a difference, which is important and provides one basis for mobilizing the humanities more broadly. They also tend to have a substantial curricular engagement, which is not surprising, as far-reaching change needs to be built around educational programs and students. However, such engagement is often made difficult by a system that privileges established disciplines and departments, especially for higher-level courses.

It is easy to understand why the fields under discussion often denounce the conventional humanities, but this denunciation can also be problematic. The challenge is admittedly structural in that the conventional disciplines absorb a great deal of resources, faculty lines, "epistemological space," and institutional leverage. There is good reason to rethink these priorities. However, the "new" upcoming fields would seem to need the humanities—and vice versa—and there is value in the interplay between newer areas and the established disciplines. And although the conventional humanities are often presented as monolithic and static, they are more dynamic than they may seem. Furthermore, specialized (or singular) studies carried out by individual researchers (or collaboratively) can make important contributions to emerging fields as well as to their own disciplines. Building capacity along the lines proposed in this book requires both.

Emerging Forms of Humanities

I use the term "emerging humanities" to describe humanities varieties such as urban humanities. A somewhat fuller list includes environmental, digital, urban, energy, medical (or health), legal, desert, water (or hydro or blue), paleoenvironmental, food, and neuro humanities.[50] These varieties are sometimes called "new humanities," although many of them have a rather long history.[51] For instance, while "digital humanities" as a name goes back to the early 2000s, the field has a much longer (and reasonably established although not uncontested) history under "humanities computing."[52] Changing the name was a move that shook up and reinvigorated the field and eventually contributed to taking it in a new direction. I take these various names, whether new or entrenched, to be meaningful to the people associated with the fields as well as having institutional power, in some case more than others. When there are several concurrent or historical labels, such as "humanities computing" and "digital humanities," they cannot be conflated, at least not without critical discussion. Similarly, "health humanities" is not the same thing as "medical humanities."[53] This can be a matter of institutional politics and programmatic direction too. For example, *The Edinburgh Companion to the Critical Medical Humanities* does not mention "health humanities" at all in the actual text, and in a chapter elsewhere on transversal humanities (and ultimately critical posthumanities), that same volume (*Critical Medical Humanities*) is referenced to substantiate a denomination that is not actually used there ("biomedical humanities").[54] Adding to this dynamic situation is the fact that these areas seem to retain their newness or emerging nature for a long time. And even with newer formations, such as "geohumanities" (with a decade-long history), there is a tendency to shorten the history of the field as if it *just* came into being ("last 2–3 years").[55]

Adding "humanities" to some descriptor to claim a new field is a productive pattern, as indicated by the above list and by new formations such as "ice humanities."[56] A recent edited volume on this topic is presented as setting out

> **a highly compelling claim for the emerging field of ice humanities, convincingly demonstrating that the centrality of ice in human and non-human life is now impossible to ignore.**[57]

As this example demonstrates, arguing for a new ("emerging") humanistic field through adding "humanities" is clearly a more aspirational move than working within the frames of a research group, specialization, program, or subfield.

New directions can also be staked out at the intersection of existing fields and perspectives within these humanities fields. For example, a 2022 handbook on medical-environmental humanities is said to bring together "two parallel and occasionally intersecting disciplines" (medical humanities and environmental humanities).[58] It is described as "field-defining" in revealing "our ecological predicament to be a simultaneous threat to human health."[59] Such descriptions point to the copiousness of the issues at hand and how the sense of acuity and emergence is regenerative. This regenerative force can lead away from existing fields toward something new or modified, such as "medical-environmental humanities," or "digital environmental humanities."[60]

Integrative, Experimental, and Posthuman

The term "experimental humanities" is sometimes used to describe these areas collectively but also more specifically to relate to experimental forms and practices in and through the humanities.[61] Another related term is "integrative humanities," which is typically used collectively but also to describe an integrative tendency driven by a realization that societies and the academy need to change:

> **"Integrative" in this discussion stands for, at the same time, interdisciplinary combinations of knowledge areas, integration of teaching with research, and a forceful and multifaceted integration of third**

mission/collaboration efforts into the everyday lives of university working departments, centres, or institutes.[62]

There is a forward aspiration in the proclamation of "integrative humanities" as part of a broader analysis of the humanities and their future societal role. The term is used analytically and programmatically at the same time. The same can be said about "critical posthumanities" as an intellectual project and framework as articulated by philosopher Rosi Braidotti.[63] Critical posthumanities is a far-reaching vision of what the humanities can be in a posthuman world beyond liberal humanism:

> **My argument is that today the critical posthumanities are emerging as post-disciplinary discursive fronts not only around the edges of the classical disciplines but also as offshoots of the established "studies." They provide the answer to what the humanities can become, in the posthuman era, after the decline of the primacy of universalist "Man" and of supremacist *Anthropos*.**[64]

Critical posthumanities can be seen as an emerging form of humanities, although operating on an aggregate or supra level. It similarly distances itself from conventional humanities and is said to "seldom coincide with the traditional humanities disciplines."[65] The new fields are collectively seen as drivers of change, serving as an "epistemological vehicle."[66] They are given considerable agency in the project, just like the disciplines invoked in the integrative humanities.

Yet another term, "generative humanities," emphasizes the production of new types of knowledge, in particular prototyping, testing, and the iterative generation of new responses.[67] These terms—"integrative," "experimental," "new," "generative," and also "hybrid"—speak to the way the fields in question are framed. I use "emerging humanities" as an analytical term to emphasize the emergent nature of these fields. They are characterized by the process of emerging, of becoming integrative, interdisciplinary, experimental, posthuman, and radical as part of a proximate future.[68] Their emergent nature gives them an aura of ongoing change, responsiveness, and being dynamic. There is often an unabashed forward trajectory (optimism, hope) in the framing of these disciplines, which is connected to institutional aspirations, a sense of urgency, and dissatisfaction with the humanities and the university as an institution.

This trajectory is not unproblematic. One challenge is to deliver what is promised and to contribute to academic and societal change for real. Extending the repertoire of humanistic work, shifting logics, and moving from talking about working with outside partners to actually doing it is a major undertaking. There is also a challenge in moving beyond setting up the conventional humanities as an opposition or a straw man to developing a humanities-driven institution outside that paradigm. Nevertheless, the emerging humanities are powerful in that they look ahead and project significant opportunities and results, ideally in both critical and constructive terms. They also serve as an important testing ground and a platform for attracting competence and building capacity.

Additionally, the emerging humanities are emergent in the sense of being regarded as being at an early phase of development, especially when compared to conventional humanities disciplines. They are not fully formed disciplines and will perhaps never be or want to be. Indeed, it could be argued that becoming too institutional, disciplinary, or entrenched would take some of their core qualities away:

> **The greatest imminent challenge facing the emerging urban humanities is how it can be absorbed into the institutional setting of the university without becoming just one more programmatic emphasis in a cross-disciplinary curriculum, or even a new sub-discipline in its own right.[69]**

This challenge is not only imminent but also substantial, and it plays out differently for different areas or initiatives. A general observation, however, is that it is very difficult to find or create institutional models—inside or outside the disciplinary template—that sustainably and humanely support in-between, collaborative, interdisciplinary work as well as a full academic range (teaching, research) and career opportunities across competences.

Positions and Positioning (Tame or Feral)

There are different interests within any scholarly community, which is not unexpected or particular to the fields in question, as can be seen in work on African-American studies and cultural studies.[70] Such interests may have different positions not only on the foundation and framing of the enterprise but also concerning other factors. For example, being more disciplinary often makes it easier to create humane long-term conditions through career pathways, teaching opportunities, and access to more sustainable funding. Institutional logic is normative, and it is simply easier to be part of the system rather than to resist it long term. The field of digital humanities is an example of a field that has clearly become more disciplinary since the late 2010s.

The emerging humanities are not all the same, and they are not internally homogenous or historically stable. For example, medical humanities were mainly associated with the professional context of medical schools but have increasingly also come to incorporate humanities-centered efforts and efforts outside medical schools, sometimes called "critical medical humanities" or "health humanities."[71] The same development can be seen with other formations, including "critical digital humanities" and "critical disability studies."

With medical humanities, there has been a worry that the field is too closely and instrumentally aligned with biomedicine and that the expansion of the field has been dependent on a reduced version of the humanistic:

> **appearing as the domain of pleasant (but more or less inconsequential) helpmeets—lurking hopefully, poetry books in hand, at the edges of the clinical encounter's "primal scene."[72]**

This quote is described as a caricature in the article, but one "not without its truth."[73] A similar argument is made by Zoe Adams and Anna Reisman (coming from the medical side) when they argue that (critical) medical humanities need to go beyond concepts such as enrichment, empathy, wellness, and "joy" and engage with areas such as health activism, queer studies, and disability studies.[74] They admit that it is more difficult to explore thornier and larger issues—such as health inequities, race-based medicine, and institutional complicity—that require hard self-examination, but they argue that this is what we need: a critical medical humanities that "prioritizes a rhetoric of discomfort and disruption" and "catalyzes deeper and actionable change."[75] These examples clearly call for a sharper and edgier humanities and for a more direct involvement.

There is a strong reflexive element here, which is common in the emerging humanities. In this case, it concerns the negotiation between humanities knowledge and a professional science-based arena. Can humanistic knowledge shift medical practice and worldviews (and the other way around)? Should the humanities be pleasant or have more bite and resistance? This self-critical reflexive element of the emerging humanities—although sometimes self-involved—is useful:

> **The emergence of the environmental humanities is at once exciting and perplexing: exciting for its intellectual dynamism and renewed sense of urgency and relevance; perplexing for its re-mobilization of concepts that have come under pressure or erasure, most notably the *human* itself.**[76]

Jennifer Wenzel points to how the environmental humanities is not as progressive and posthuman as might be expected. For example, she argues that the field exhibits human-centeredness through its engagement with the Anthropocene. Wenzel describes this tendency as a troubling new universalism.[77] This debate, as articulated within environmental humanities and other fields, is relevant not only to the formation and framing of these fields but also more generally to infrastructure building. For instance, what might it mean to pay more attention to a fuller context, including nonhumans, in efforts to create frameworks for planetary management or better medical bots?[78] It also points to the divergent viewpoints present within the same field.

As we have seen, the emerging humanities provide a fertile ground for competence, experience, and tensions that can productively feed into building capacity for mobilizing humanistic and humane knowledge. They are also particularly relevant because they tend to have societal engagement, directly relate to infrastructure, and often include an experimental ethos. Environmental and urban humanities, for example, connect to massive social challenges and to environmental and urban infrastructures. In this sense, they are challenge driven and infrastructural. This does not mean that everyone in these areas would be suitable for or interested in humane capacity building or systematic constructive engagement but rather that there is an overall potential.

Fields Ending in "Studies"

A World of "Studies"

There is a group of fields ending in "studies" (sometimes "theory") that share some of the properties of the emerging humanities and which also seem relevant for identifying potential for humanistic and humane capacity building. Disability studies, for example, can be said to have an interest in how the world is built and experienced, how infrastructural templates can be shifted, and there is often a strong civic and activist engagement. Unlike the emerging humanities, this group includes fields not clearly placed in the humanities or identified with the humanities but also in the social sciences and more broadly.

Many of these fields emerged in or after the postwar period in conjunction with the rapid expansion of higher education as well as with central political developments such as the civil rights movement. Contemporary examples include animal studies, area studies, cultural studies, women's studies, gender studies, ethnic studies, death studies, media studies, crip theory, postcolonial studies, queer theory/studies, food studies, and disability studies. Several of these fields, including ethnic studies and queer studies, fundamentally engage with social and political issues that are tangible and current. Other fields, such as information studies and media studies, seem more institutional and less radical (and hence less important for the current study), at least at this point in time. In any case, the genealogies are exceptionally rich and varied and cannot be covered in any detail here, but there are some observations that can be made before moving on to two examples: disability studies (and some related fields) and food studies.

These fields are rich also in the sense that they tend to be self-reflexive and well documented, just like the emerging humanities. Such material can be invaluable for teasing out some of the issues at hand for humanistic capacity building. With disability studies, for example, one concern has been the balance between critical and constructive work:

> **Can an intellectual home be found (or created) where abstractions and theoretical analysis are juxtaposed with policy and practice without either claiming priority of place over the other?**[79]

A similar question is posed by critical disability studies researcher Dan Goodley, who, in advocating for an increased and more radical theorization of the field, also asks for "spaces for the development of praxis: the intertwining of activism and theory."[80] Such questions occur in the history of many of these fields. For example, gender scholar Rachel Corbman discusses how the formation of women's studies can be understood in terms of "tension" between activism and scholarship based on an analysis of the 1982 Barnard Conference on Sexuality.[81]

The interface between scholarly work and active engagement is relevant when considering efforts to instrumentalize the humanities. An important question concerns when there is obvious or general usefulness of knowledge produced by such fields but a risk that the specific origins or questions can be lost in the process or be employed irresponsibly. For example, media scholar Jacob Gaboury argues that there is a risk that the radical potential of queer studies gets watered down and perhaps also disassociated from that potential:

> **While we may desire to identify or prescribe a queer computational form that we might use to disrupt or transform our digital landscape, we must acknowledge that any such prescription opens queerness to a radical visibility such that it may be repurposed to serve a normative ideology.[82]**

Although perspectives and methods such as "glitch" can be useful in challenging and distorting norms and systems,[83] they can be appropriated not only by normative ideologies but also by careless instrumentalization. This is why it is so important to facilitate considerate and constructive conversations between actors and to build capacity around trust, sharpness, and curatorship (chapter 4).

Negotiating Frames and Genealogies for "Studies"

Because many fields ending in "studies" or "theory" are newer and more heterogeneous than long-standing established disciplines, they more often undergo periods when the field itself is discussed and debated. Just like with emerging humanities, the risks and benefits of institutionalization make up one source of debate in these fields, at least at some point in their history.[84] This question is also linked to the issue of being or not being interdisciplinary. Jack Halberstam argues that queer studies has benefited through its stealth approach (not being fully institutionalized), which has kept the field multidisciplinary ("a promiscuous rogue in a field of focused monogamists").[85] We are reminded here of Sandy Stone's strategy of operating "under the radar" for the ACTlab at University of Texas, an institutional "embodiment of Trans practices writ large."[86]

Another similarity is that these fields—just like emergent humanities fields—tend to be oppositional or have an outside position. They came into being despite or because of the established structures of academia, which, for epistemological, institutional, tactical, or political reasons, could not well enough accommodate the work and perspectives envisioned. Under one analysis, the field of queer studies has positioned itself as "constitutionally against the grain, athwart the academy."[87] Another example is the field of cultural studies, which has been described as oppositional to the basic logics of the academic enterprise, including the disciplinary organization of knowledge.[88] Neither of these fields, however, was instituted only because of such infrastructural or institutional reasons.

There is naturally an entanglement of topical and structural motivations. Queer studies came about as new or emerging ways of thinking about sexuality and as a far-reaching critique of normative models of sexuality, sex, and gender.[89] Cultural studies, on the other hand, can be seen as a response to a literary elitist tradition, which framed popular culture as a threat, as well as the manifestation of oppositional ideological direction.[90] This oppositional stance—at least at certain times in the establishment of these fields—contributes to creating particular conditions for knowledge work, including an explicit wish to challenge institutional structures and the world outside.

Rosi Braidotti extends this outside position further when she describes these fields as "creative trans-disciplinary hubs" as part of what she calls "critical posthumanities."[91] She describes first-generation "studies," such as queer studies and feminist studies, which have worked toward decentering humanistic Man through a far-reaching critique of academic humanities, their structural anthropocentrism and Eurocentrism, and through engaging with the present and taking power seriously. According to Braidotti, a later generation of "studies," including posthuman/inhuman/nonhuman studies and fat studies, builds on this genealogy but has a stronger interest in addressing the issue of anthropocentrism directly.[92]

The progressive assumption of this genealogy can be related to the work of Cary Wolfe, also on posthumanism, where he notes that the humanities and liberal humanism have tended to make pluralism, an extension of perspectives and things studied to marginalized groups, a matter of incorporation rather than "destabilizing or throwing into radical question the schema of the human who undertakes such pluralization."[93] A similar argument in relation to emerging humanities has been made by literary scholar Ortwin de Graef, who sees these new varieties as responding to "shifts in truth-and-relevance regimes by fashioning themselves to fit newly dominant frames that appear to exclude them."[94] Wolfe points to the importance of engaging both on a subject level and on the level of methodology and theory. Animal studies is not just about studying animals as a topic but also about embracing "nonhuman" and "animal" as a perspective. Human and cultural geographer Franklin Ginn argues that the objective of animal studies is topical and political.[95] The latter objective includes cultivating "new alliances for multispecies flourishing and recuperation, alliances that will be necessarily partial and incomplete."[96] What is at stake here is to what degree the topic of a field and its inflection can influence what that field is about and how it gets enacted.

There is a risk with collectivizing these "studies" areas, just as with the emerging humanities. It is true that they share some properties and are part of a productive formational pattern, but they also have very different lineages and dependencies. While representatives for various fields tend to valorize their own emerging field and its potential—which can and should be discussed critically—analytical and programmatic approaches to these fields as a group can similarly ascribe them collective qualities that also must be discussed critically. This is true of the narratives of "integrative humanities" and "critical posthumanities" as well as "emerging humanities." For instance, the search for potentiality in this chapter and Braidotti's identification of transdisciplinary hubs of importance are seemingly worthwhile (and overlapping) goals, but these need to be explored in a hedged, nuanced, and reflexive way.

The progressive and radical framing of "studies" and their emergence, for example, can contrast with actual historical trajectories. An example is area studies, which in the US can be traced back to government intelligence needs in the Cold War era.[97]

> **Did the humanists really subvert the aims of their funding sources, or manage to ignore them, or actually fulfill an enlightened construction of those aims?[98]**

This is not to say that area studies has not produced a wealth of important radical work but rather that it lives (lived?) within this frame and historicity. Digital humanities, which is now often taken to be a progressive critical enterprise, comes out of a strongly technological and instrumental tradition.[99] Again, this does not mean that digital humanities are not progressive and radical—they are so much more than earlier—but there is considerable complexity to the argument, including the tensions and inherent worth associated with the instrumental-technological tradition. There may also be resistance within fields to the programmatic visionary role given to them from both the inside and the outside.

Disability Studies

Let us now look at disability studies, crip theory, and crip technoscience as an example while recognizing their differences. These would seem to have a clear societal and infrastructural engagement that also involves some type of constructive or design ambition:

> **This manifesto calls attention to the powerful, messy, non-innocent, contradictory, and nevertheless crucial work of crip technoscience: practices of critique, alteration, and reinvention of our material-discursive world. Disabled people are experts and designers of everyday life. But we also harness technoscience for political action, refusing to comply with demands to cure, fix, or eliminate disability.[100]**

It seems clear that the space opened by disability studies scholars Aimi Hamraie and Kelly Fritsch (and others) under the rubric of "crip technoscience," going back to a complex genealogy including disability activism and environmental justice, is one of active humanism and combined critical and constructive grounding.[101]

An example of applying this type of framework comes from M. Remi Yergeau's critique of hackathons—data-intensive sessions to "hack" solutions to various problems, which they describe as a new form of telethons.[102] Telethons are fundraising events for charity, described by Yergeau as a model dependent on the pitiable and hopeless poster child, the celebrity spokesperson, and promising a cure that needs to be funded. In the hackathon format, which is ultimately about using and building infrastructure, the cure is typically an app or another technological solution or cure. The disabled community is often not even involved in the process.

Yergeau traces three lineages for disability hacking: hacking as passing (feigning normalcy), fixing (disability is a puzzle to be solved, bodies need to be cured), and retrofitting (adding to an already existing design).[103] These can be applied to infrastructure thinking more generally. There is typically a strong template (normalcy), a problem to fix, and when an infrastructure needs to be adjusted, it is usually amended within the infrastructural framework in question rather than outside it.

Yergeau suggests a criptastic variety of hacking that fights against forced normalization as well as associated frames and that is participatory, messy, and dynamic rather than static, additive, or clinical. Hacking is thus conceived through how bodies and spaces move and interact—process—rather than through apps and techno-cures. Yergeau argues for a shift in logic on several levels, in the way that the "problem" is conceived but also in how it is approached. They argue that the disabled community needs to take the lead.

In her book *What Can a Body Do?* disability and design researcher Sara Hendren revisits how she—as an artist and writer—used to think of engineering and technology as straightforward building and about how fast and efficiently things could run.[104] She admits that she earlier regarded engineering as a limited way of seeing the world, not capable of responding to the large issues she thought was most important. In her work at Olin College, she came to revise her position. A leading example in the book is about the collaboration between a class of engineering students and Amanda Cachia, a curator and scholar who is four foot three (1.30 m) tall. Hendren functioned as the teacher or convener. She describes how the students came into the classroom with a positive constructive mindset but also mainly focused on technology and solutions:

> **She wasn't looking to be fixed—not in the least. . . . We were invited to design an object that fully captured all the energy of our imagination—something between a tool and a piece of architecture that roundly rejected the standard dimensions on which the built world is generally arrayed, and proposed instead a way for Amanda's body and the room to meet each other differently.[105]**

The process facilitated and described by Hendren is quite different from conventional design processes and one that aims at revising and shifting logics rather than accepting or adjusting a given frame. For example, the main goal is not to empathize with users, as is a common strategy in so-called design thinking and human–computer interaction. Rather, Hendren's approach resists such a model. In fact, her work can be seen as an implicit response to empathy-driven design, which has been critiqued for sustaining the differences it seeks to elide, missing out on key concerns for the users that do not fit the design process, and not being responsible or accountable.[106] It can also be seen as a reaction to so-called universal design as an all-for-one approach—a critique that has also been voiced by other scholars.[107]

Design and art play a foundational role in Hendren's work, together with the broader context of disability studies. She reflects on how she was looking everywhere for "examples of disability representation that had both the workhorse pragmatism of good design and the complexity of culture—expressive language and materials, a sense of dimensional personhood, strong poetry."[108] Amanda Cachia, from a related but different perspective, challenges conceptions through asking questions along the lines of what would happen if artists and curators made access into a critical integral part of their artwork and curatorial practice, considering it as a "theoretical and practical generative force."[109] While disability studies as a heterogenous and complex field rarely manifests the full implementational ethos of Hendren's and Cachia's work, it clearly demonstrates engagement:

> **The handbook represents the tensions between academic scholarship and the passion of activists; the different perspectives of disability studies and rehabilitation sciences; the uneasy coalition of disabled people with health professionals, technicians, and policymakers; and the value conflicts between aggressive capitalism, social welfare states, and the poor who struggle for survival in industrial and developing countries.[110]**

The tension, uneasiness and variety of positions are part of "disability studies," just like the notion that the field proposes a different way of looking at disability. Dan Goodley says that the field brings together theory, research, practice, and activism that contest the idea that disability equals human failing.[111] These are logic-shifting and multi-perspectival concerns that, together with the "making" sentiment of some parts of the field, are deeply informative for humane capacity building.

Food Studies

Some implementations of food studies, especially ones focused on culinary or nutritional aspects, seem worlds away from a field such as disability studies. But in a broader context, the field of food studies shows similarities to values- and challenge-driven fields. Food consumption is culturally, historically, and politically embedded. Food security, climate mitigation, and sustainable food systems are key global challenges. The food studies field also has a clear infrastructural and material aspect to it—think of food supply chains, urban farming, or shifting away from meat consumption. One institutional account describes food studies as a

> **burgeoning, interdisciplinary, inherently politicized field of scholarship, practice, and art that examines the relationship between food and all aspects of the human experience.[112]**

This description points to several of the features we associate with challenge-driven and values-driven fields, including the emergent nature of the field and the interdisciplinary approach used. There is also a future-oriented component to the description, as students are said to be expected to develop ideas about how they can "contribute to solutions and explore multiple pathways to transformation."[113] It should be noted that the field of (critical) food studies is one field among many often-overlapping areas involved in investigating and shifting food systems.[114]

A 2021 design and food studies event at the Norwegian University of Science and Technology (NTNU) can serve as an illuminating example of the landscape of interdisciplinary food studies. The panel was described as discussing "ethical and sustainable food futures!"[115] Most of the participants were early-career scholars, which is perhaps telling. Several identify with food studies or critical food studies in their work. The work discussed focused on sustainable cities, food systems and social movements (cultural anthropology), novel sustainable proteins including edible insects (design, sustainability studies), poultry farming in the city (design), the mental social and physical ecologies of urban gardening (architecture), and sustainable meat use in Norwegian food practices for climate mitigation (science and technology studies). There was also participation from a local NGO supporting newly arrived refugees through organic farming and an organization keeping honeybees as an educational tool. Such platforms would seem to be natural allies for building capacity for humane infrastructure building, and the connections to urban and environmental humanities as well as to urban and environmental infrastructures are suggestive.

The sustainable meat project presented at the NTNU event is called "MEATigation." It sets out to explore how meat is part of Norwegian food practices and attempts to find ways to promote sustainable meat use in the country (and more broadly).[116] The basic assumption is that Western meat consumption needs to be reduced in order to respect planetary boundaries, ensure global food security, and improve human and animal health. Even when people are positive to taking climate action, there is a reluctance to reduce meat consumption (not least in Europe and the Nordic countries), and the project attributes this to the fact that food is culture as well as calories. According to the project description, MEATigation investigates how social practices, such as eating or providing meat, can be guided or shifted by, for example, recognizing the people and animals working to produce meat or by using plant-based, insect-based, or in vitro alternatives instead of animal-based meat.[117] Recipes present one opportunity for remixing and shifting frames. Project lead Sophia Efstathiou, a philosophy of science scholar with a background in science, refers to recipes as "experimental protocols or algorithms."[118] There is a playful and experimental ethos to the project, and Efstathiou emphasizes the importance of artistic practice as well as conventional scholarly modalities—in fact, bringing them together.[119]

MEATigation received funding from the Norwegian Research Council through the KLIMAFORSK program (2020–2024). It is thus an example of a food studies–oriented research program funded within the context of climate change. The funding program is challenge focused and interdisciplinary and, interestingly, has stressed the importance of humanities and social science research in the program, not just nominally but as a proportion of the project portfolio (and funding allocated).[120] Nevertheless, the program is unsurprisingly mainly science based, and according to a 2021 evaluation, the social sciences are in a much stronger position than the humanities (31 percent vs. 3 percent of the total funding).[121] Also, the project portfolio consists mainly of projects from a single discipline area, such as "mathematics and natural sciences" or "social sciences" (72 percent in 2020).[122] The way such programs are set up—a kind of conditioning—is relevant for areas such as food studies but also more generally for building humane and humanistic capacity.

Policy expert Julia Stamm points to how the gap between rhetoric and practice in science policy needs to be reduced, for example in relation to challenges such as climate action and food security.[123] It is well known that despite an increased emphasis on the role of social sciences and humanities in relation to societal challenges, for example in the European Union Horizon 2020 program, such goals are challenging to implement in practice, especially securing a more active role for these areas and seeing them as full partners rather than them being given or taking a service function (e.g., to help make technological solutions become accepted and understood) or an indirect role (providing embedding, ethics, detached research, or an alibi).[124] There are also expectations that such interdisciplinary efforts will just happen rather than being processual and in need of nurturing.[125]

In this light, MEATigation—as a project driven by social science and the humanities—represents an important and meaningful development, especially as it is unapologetic, processual, and outward facing. We need research funding and science policy that create conditions for such and other similar efforts, especially in academic environments highly reliant on external funding. In the case of frameworks such as KLIMAFORSK and Horizon 2020, one challenge moving forward is to more strongly support initiatives driven by several knowledge areas—for example, humanities/social sciences and science/technology—at the same time and from the very beginning as well as to consider capacity building across projects. Another important factor is how funding and support schemes are actually implemented. For example, MEATigation received a grant of NOK 14.8 million (USD about 1.5 million), which is substantially more than is typically given to social science and humanities research projects.[126] It is also telling (and encouraging) that most of the research team consists of early- or mid-career scholars and scientists.

Interlude: "I Had to Ask Myself Why I Wasn't Working on Climate"

The invocation of new vibrant fields and exciting collaborations is at least implicitly based on the idea of energetic and engaged participants, in particular incoming early- and mid-career faculty and experts as well as students. They are integral to the emergence and associated hope present in those institutional stories. Conversely, it can be argued that people come to these fields because of stories that embed interdisciplinarity, social engagement, support of multiple modes of engagement, and the promise of intellectual and social engagement coming together.

Such narratives come with responsibility. The institutional work needs to be caring and responsible in relation to these people as well as sensitive to the conditions of alternative scholarly work and platforms, including the challenges of operating across epistemic traditions, managing nontraditional career paths, and securing accreditation for different types of knowledge production. There also needs to be a discussion of how to balance intellectual and social engagement and how to navigate a territory ripe with hope and dread.

Individuals have trajectories just like institutions and fields, and there are multiple pathways to the sites and forms of knowledge production under discussion in this chapter. One pertinent question for the individual can be how to reconcile research, teaching, and an active or activist engagement in relation to challenges in the world. Another common issue concerns being in several places at the same time, as people often end up being affiliated to a new initiative as well as a conventional discipline. This is not necessarily a bad thing but requires individuals to negotiate different epistemic traditions and institutional aspirations. There are multiple legacies, dependencies, and goals in play.

Some of these complexities came up in a text-based conversation on "Changing Ourselves—Climate, Carbon and Concepts" between media technology scholar Daniel Pargman and media studies scholar Anne Pasek that I convened in 2021.[127] Pargman is based at the KTH Royal Institute of Technology in Stockholm and Pasek at Trent University in Canada. I invited them because they are doing inspiring work relevant for humane capacity building. Both also shifted their academic trajectories as a result of things going on in the world, in particular environmental concerns.

The conversation was asynchronous and was meant to be informal and open, more or less like a spoken conversation.[128] A selection of the conversation is presented below as two aggregated blocks of text, hence sacrificing the flow of the conversation but hopefully maintaining the flavor of the whole exchange:

> **Anne Pasek:** I have a background in art and art history, and I initially wanted to do my PhD on the history of computer failure and all the complicated emotional and aesthetic ways the frequent non-functionality of computers gets negotiated in design and in cultures at large.
>
> I was all ready to start that project and had won a Fulbright grant to do so in the US when I was scooped up by Fulbright, along with all the other foreign grantees, and plunked in LA for a weekend. This is a regular part of the experience; they want to foster some networking for participants, but they also want to show off American policy solutions by bringing everyone together to talk

about a current issue. Well, our issue was the drought that was happening in California, and their treatment of the problem had the opposite effect. We were shown a lot of genial nonprofits working on water conservation plans for the private single-family homeowners who cared about the issue, but larger structural problems were never discussed, and I don't believe anyone even mentioned the words "climate change." Suffice to say, it did not inspire confidence. It actually got me rather anxious, puncturing the belief that there were a host of very competent people out there doing climate work throughout the academy. Instead, I had to ask myself why I wasn't working on climate, given the stakes and gaps I was seeing even at the heart of this supposedly very prestigious and powerful nexus of scholarship.

I like to say that instead of studying how we deal with computer failure, I now study how we deal with climate failure and communication failure. I'm interested in where barriers keep coming up. In part, I'm working on a book on the ways we've tried and often failed to make "carbon" into a socially meaningful concept. I'm also working on another book about the ways that Moore's law and energy efficiency helped make computing seem like a "clean" or "immaterial" industry and the ways we might shift that understanding. And I'm also working to coordinate a wider research group, thinking about the ways in which the social norms, institutions, and methodological protocols in many corners of academia lock us into high-carbon rhythms, with implications for the kinds of knowledge and researchers these systems make possible (air travel is such a large part of this!). We're exploring how low-carbon research methods might therefore offer both epistemological and equity gains or at least provide a very interesting vantage point from which to think about systems of knowledge production.

On my end, I've gotten pretty good at compartmentalization. I think it's important to pause and feel those feelings of grief and loss and political frustration, but I find it the most useful to spend most of my time working with a sense of motivating anger and urgency. I also try to have a few projects on the go that "fill up my cup" rather than solely drain it. I sometimes wonder if this is the real use of art in climate communications: not necessarily communicate a message to an external public but to provide a space for us to think through the implications and challenges in front of us with different parts of our brains. For me, that's meant making gloomy zines about my hometown's attachment to oil, funny videos for climate organizing work, and a solar-powered server in my backyard. These things aren't commensurate with the scale of the problem, but they are a way of staying with the trouble, at least.

Daniel Pargman: I wrote my PhD twenty years ago about virtual communities and online meeting places: "Code Begets Community: On Social and Technical Aspects of Managing a Virtual Community." When I was a guest researcher at Microsoft Research in 2007 (and living in car-dependent suburbia US), I had an "environmental awakening" and realized that nothing that I was or had been interested in (including the research that took me to the US) was really important on the larger whole. I then spent years reading widely, but mostly nonacademic texts (books, blogs, etc.) about peak oil, energy, and sustainability. At that time, I couldn't see a strong connection between sustainability and my day job. So, reading up on energy and sustainability was akin to being a "hobby" of mine that I pursued manically. Five years later, I saw such connections, and I have, during the better part of a decade now, pivoted toward sustainability professionally.

Together with a colleague of mine, I now lead a research group at KTH Royal Institute of Technology (Stockholm), and we are running many different exciting research projects. We also teach about computing and sustainability and will give our flagship course for the tenth time this autumn. The one research project with the best fit to the stated framing of this particular conversation is a project where we aim to support the climate targets for reducing CO_2 emissions from flying at our own university. The image below shows all employees' all flying at a specific department (2019 data, figure 3.1). Each poker chip represents a one-way trip. Green chips are short-haul trips, red are medium-haul trips, and black are long-haul (intercontinental) trips.

Another thread we might discuss that both Patrik and Anne touched upon above is how to manage the interface between the private and the professional when we are engaged in issues we think are really important (sustainability) both at and outside of our work. What is sensible, and how do I get the most "bang for the buck" if I want to change the world? Should I aim for being the most excellent researcher I can be? Or the most charismatic and persuasive teacher there is? Or the most passionate and knowledgeable activist around the block?

I have, over time, come to the conclusion that it could be that my largest impact is through my teaching, where every year I reach eighty to ninety students who will work in industry for decades, rather than (for example)

Figure 3.1
Approximate mock-up based on the original flights visualization from the KTH project.

> through my own personal actions in my private life or the on-the-ground actual impact of my research. That ought to mean that I have a responsibility (towards myself) to do the very best that I can when I teach. That obviously means going beyond "transferring facts" to actually connecting with my students as well as making them understand some of the conundrums and trade-offs we face (there are no easy solutions ahead). One challenge we face is that our teaching can evoke strong emotions among our students. We don't shy away from that, but I have to admit that we, on the other hand, also don't exactly know how to handle this. Handling students' emotions is not something that researchers/university teachers traditionally are trained to do, but we have learned along the way (our sustainability course will be taught for the tenth year at the end of 2021).

Even if both Pargman and Pasek shifted their academic trajectories because of their grave concerns about the environment and climate change, they also hold on to those same trajectories, where they came from. They shifted topics and orientation but built on their disciplinary backgrounds and knowledge. Pasek earlier worked within a media studies department and is now affiliated with both the Department of Cultural Studies and the School of the Environment at her new university. Pargman was already based at the Division of Media Technology and Interaction Design but changed his academic focus within that organization. It should be noted that media technology and interaction design is probably one of the least "engineering-like" of the technological programs at KTH and so is perhaps more amenable to new orientations.

In another part of the conversation, Pargman talks about papers on teaching he and his colleague (Elina Eriksson) have published almost every year, including "'It's Not Fair!': Making Students Engage in Sustainability" and "On the Inherent Contradictions of Teaching Sustainability at a Technical University."[129] A broader group involved in the program has also published more comprehensibly on eco anxiety.[130] These titles point to the important institutional work they carry out as teachers to change the curriculum in the context of an engineering school, where the potential impact for long-term societal change is substantial but also exemplifies the tension that Pargman references in the excerpt. Both Pargman and Pasek point to how it can be difficult (and is important) to handle the emotional investment by students. Pasek describes how she uses first-person commentary to make "having feelings" less exceptional and how she attempts to open up a space in the classroom that can accommodate dialogue across positions in her diverse student groups. This work is about not just the students but also the teachers, and about not just managing anxiety or emotions but also responding to students who are invested in change in relation to their disciplines and future professions and more broadly.

The conversationalists are driven by strong personal convictions in their academic work, but there is also a separation between personal engagement and the academic side. In the excerpt, Pasek talks about compartmentalization and not letting grief, loss, and political frustration dictate her work. Pargman talks elsewhere in the conversation about how he is "quite cynical on the whole" but also how he does not "for the most part let that affect my day-to-day outlook." He works toward the Paris Agreement every day but finds it increasingly hard to imagine that we will be successful. Pargman describes how climate change became an obsessive hobby for him, and Pasek points to how she engages in projects on the side that are relevant but also not directly academic. These would seem to call on modes of engagement that she may not readily make use of within academia (e.g., making zines, funny videos, and a solar-powered server). These are real, of course, but also representational acts.

The positionalities of the interlocutors are important. Pasek comes from a critical studies perspective with a focus on deconstruction and Pargman from a "soft" critical engineering background, and these epistemic positions are evident in the conversation as well as through how they are being negotiated and challenged. Pasek is more engaged with critiquing the underlying conceptual systems, for example carbon footprint as a framework and

Moore's law as an idea, whereas Pargman seems more focused on maximizing the operational logic of existing systems. Both positions appear to enable and prevent certain types of action and dialogue. While Pasek's critical deconstructionist perspective provides a broad scope for new ways of thinking about social reconfiguration, Pargman's view offers a narrower but potentially more operationizable vision.

Both Pargman and Pasek work in a textual tradition where publishing papers and books make up a key goal, but it would seem that Pargman has access to a different and perhaps more extensive toolset (academically) because of the disciplinary focus on information and communication technologies, including visualization and technology-driven "futuring." On the other hand, he seems more reluctant to take an activist stance academically. He delivers the data from their academic flying project, and he points to Agenda 2030, but he does not readily suggest how to move forward (essentially leaving that to university management and policymakers). The markers representing university flying are suggestive by themselves though, and it is telling that Pasek and Pargman come together in their reflecting on their own sector's infrastructures, associated practices, and the question of complicity. For Pasek, this "smaller" focus also seems to bring forth a stronger element of direct, experimental, and action-focused engagement and to drive program building. For example, her recent work on low-carbon research includes white papers on conferencing and managing carbon emission anxiety through infrastructural experiments and is part of a programmatic effort led by Pasek.[131]

The conversation between Pasek and Pargman exemplifies the genuine engagement, tensions, epistemic nuances, and program building that can emerge in a collaborative space around challenges. The professional, institutional, and personal investment they demonstrate is inspirational and important.

Instrumentalization, Instruments, and Toolboxes

Arguing Usefulness

The humanities have a tendentious relationship with concepts such as "usefulness" and "instrumentality," and they tend to be reluctant to frame themselves actively (chapter 2). There is an evasiveness to the field with a long history.[132] We will now consider some of these tensions as we move toward creating toolsets and building capacity.

The humanities are often passively or indirectly framed, both by humanists and by the outside. When the humanities engage, they often get or take on a peripheral, subservient, or ornamental position in relation to "real-world" projects. While this framing does not always inhibit active engaged work on the individual or group level, it can make humanistic and humane capacity building difficult. Furthermore, the humanities tend to be resistant to capacity building, hesitant to engage with innovation-driven agendas or "infrastructure speak," and skeptical about top-down attempts to control their research agendas.

This skepticism is often warranted. There is a risk that the rich criticality, historicity, and contextuality of humanistic work get watered down or lost if transferred into ideation, innovation, and making (an entrepreneurship model). Think of when "ethics" gets added to various technology-driven initiatives, when the humanities are understood as language or communication skills, when design gets redressed through adding "thinking," when humanists who attempt to tackle real-world challenges or build things get caught up in the technological or instrumental side of things, or when humanities-oriented researchers at social media companies are ousted.

In a response to a push for research funding to be driven more strongly by societal challenges, the Scientific Council for Humanities and Social Sciences at the Swedish Research Council argued that they were already meeting this need and that it makes more sense to increase investments in their regular funding schemes.[133] The main argument put forward in the report is that humanities and social science scholars are already engaging with these themes without the need to orchestrate such research from the above. However, an important question is what type of work is supported through such grants and whether bottom-up facilitation through conventional funding mechanisms can build capacity to respond to societal challenges.

The council used an extensive frequency analysis and topical modeling of proposal abstracts to support their argument.[134] The data show some increase for most topics they looked at (climate, democracy, gender, migration, mental health) between 2010 and 2017.[135] However, the data are inconclusive, and in some cases, an increase over one or two years is followed by a decrease. One problem is that there is no systematic comparison of these numbers and trends in relation to other topics. A parallel analysis—of abstracts from specific research committees—shows that many other topics are much more common than the topics associated with challenges.[136]

Overall, there is no sense of a surge or persistently increasing engagement in challenge-driven research in the humanities and social sciences in the material presented by the Swedish Research Council in the report. There is also no discussion of what it really means that project abstracts (not the actual proposals) contain "challenge words" within a funding category that mostly supports individual or small projects within one discipline. It would seem, for example, that this frame suggests certain modes of engagement and types of scholarship.

A related type of situation is when there is a strategic initiative at the level of an institution, and the humanities negotiate their participation, either before or after the fact. An illustrative example is provided by a Yale University initiative announced in early 2024 to "explore the transformative possibilities of AI at Yale."[137] This can be seen as a repeat effort following many other initiatives at other institutions, and it largely follows the template established. The high-level task force announced, for example, almost exclusively consists of technologists, scientists, and social scientists.[138] A week after the announcement, two Yale faculty published an opinion piece in the *Yale Daily News*, where they argue for the importance of a critical humanities perspective on AI, partly drawing on their work on an undergraduate certificate in critical computing studies.[139] Their criticality argument is profound and convincing. However, the presentation remains in a deconstructive modality, portraying the humanities as an outside actor with mainly critical powers, which is particularly noticeable toward the end of the piece:

> **Academic-industry collaboration can have productive synergies if we avoid mission creep. . . . We admit that we critical humanists sometimes earn our reputation as troublemakers. Before we go straight to the solution, we want to know what the problem is, how it is defined, and by whom. It can feel insulting to problem-solvers and system-builders to hear their language might not be totally under their control or their taken-for-granted ideas carry implications, sometimes dangerous ones, that stretch far beyond their labs and design screens. . . . Interpreting basic concepts, understanding deep historical legacies, asking about possibilities, and probing power—these are the comparative advantages that critical humanists can bring to the conversation about artificial intelligence at Yale.[140]**

Their position is essential and well articulated, especially given the context of a brief response in a charged situation. What seems to be missing, however, is a "how," in particular in relation to how the described criticality can be integrated into the AI efforts at Yale. Such a direction probably requires modeling the humanities beyond their (all-important) critical deconstructive powers. For example, there is no wish expressed in the piece to be involved in *formulating* the problem and the "solution." In this case, there might not have been an opportunity, but as a general principle, it would seem important for the humanities not to come to the table late when the frame has already been set. Critical scholars should definitely participate as troublemakers, but as suggested in this book, they can also be part of building systems and "solving" problems (especially if they have contributed to the framing of both problem and solution).[141] Such engagement needs to be critical and constructive and requires humanists to acknowledge and even embrace their complicity. It also requires the institution to engage humanists at an early stage in substantial ways, which does not appear to be the case with the Yale AI task force. Such processes are two-way streets, however, and it seems unwise to overemphasize the separation between the humanities and "the others," for example through implicating that the non-humanists would not be attentive to implications "beyond their labs and design screens" or reinforcing positions through invoking "mission creep" as a corporate strategy (mission creep can go both ways and is not always a bad thing).[142] The point is not to make corporations benevolent or label all "others" as critical intellectuals but rather to be respectful and find middle ground (which is a responsibility for everyone involved).

It is important to stress that the humanities and social sciences have contributed importantly to the shaping of our societies and continue to do so in multiple ways, including through providing rich and important scholarship, undergraduate and graduate education, and public intellectuals. This role has often been encoded through certain keywords, such as analytical skills, reading, culture, and values:

> **For centuries, the humanities were where cultural values and understandings were developed and debated; where history was uncovered; where heritages were preserved and assessed; where abstract thought was allowed to roam free.[143]**

Looking for impact and importance of the humanities outside the university has consequently tended to include factors such as overall "consumption" of (high) culture, the vibrancy of cultural debate, and the well-being of a nation (often citing the negative polarities of these) as well as educating new generations of citizens. It should not be surprising, however, that using indicators such as reduced time spent reading for personal interest, the reduction of the space for literary and cultural debate in newspapers, or the prevalence of references to "humanities" in obituaries does not add much strength to an institutional argument for the humanities.[144]

There has recently been an extended discussion of the impact and use of the humanities in the light of increased emphasis on employability and innovation in higher education, partly as a valuable effort to reframe notions such as impact and innovation.[145] Humanistic impact indicators brought in by such work include education and alumni, circulation of knowledge, policy influence, public talks, the role of the humanities in the media, and influence on government official reports, education, and alumni.

It can be argued that the humanities need to look inward as well as outward and that cultural shifts and a rethinking of credentialing mechanisms are required. Such work includes the way that teaching sessions are set up, as in this inspirational instruction under "class credo" for the course "At the Interface" taught by media studies scholar Jonathan Sterne:

> **It is everyone's job to imagine a better world: any critique of how something is raises the question of how it ought to be. You will be asked to think carefully and imaginatively about alternatives to the way things are.**[146]

Other efforts focus on the internal evaluation of nontraditional modes of engagement, such as the "Guidelines for Evaluating Publicly Engaged Humanities Scholarship in Language and Literature Programs" developed by the Modern Language Association, which usefully acknowledges a diversity of outputs and the complexity and reciprocity of collaboration.[147]

Instrumentalization of the Humanities: Looking Back and Forward

An example of outward engagement often invoked in Swedish historical research is government official reports and parliamentary and state committees, which reflect a long-standing tradition of government using humanities and social sciences researchers as experts to look at different societal phenomena as well as the emergence of an "expert culture" in the 1930s and 1940s driven by the project of rationalization.[148] These scholars often had strong positions in such committees and helped provide a basis for political decisions through the resultant, often extensive, reports. It is perhaps telling that this paradigmatic example is one where the primary knowledge product is a printed report or book. Although the reports were different from regular academic products, they align well with the epistemic tradition of the humanities and social sciences.

Having a connection between the spheres of higher education and government was seen as productive in this development, but it had to be carefully conceived and was generally seen as a one-way street from research to policy, described by sociologist Per Wisselgren as a one-way cat-flap door.[149] The political scene changed in the 1980s, however, and the kind of extensive and deliberated process invested in knowledge production that had been common was shifted as a result of several factors, including changing conditions for policymaking and the instrumentalization of higher education research.[150] One consequence of this shift is that there is likely less space for boundary work in Swedish academia than earlier, especially in the humanities and social sciences.[151] One might perhaps also ask if there is a risk that the partly manufactured polarization around issues such as identity politics and cancel culture will further reduce this space.[152]

The current consideration of impact and innovation in the humanities and, more broadly, the human sciences is part of a longer history of legitimization and instrumental push, which can help us better understand issues related to the "use" and alignment of the humanities. As historian of science and ideas scholar Hampus Östh Gustafsson demonstrates, this history of marginalization and legitimization is longer than has often been assumed, going back to the interwar period in Sweden.[153] For example, the focus on ideals from an aristocratic society in the humanities was not necessarily compatible with the introduction of political democracy and the emergence of the Swedish welfare state in the 1930s and 1940s.[154] Higher education was increasingly expected to contribute progressively to changing society as it became a democratic rather than an elite project. Knowledge politics were getting shaped by notions of efficiency, rationalism, and planning.[155] Over time, universities became a sector for political reform where technological rationality played a central role.[156] This development not only put pressure on the humanities but also made the humanities respond in different ways, both in accommodating and in resistive ways.

An accommodating response from the humanities that emerged in the mid-1950s in Sweden was to align with rational planning, for example through construing the field as a balancing force that could compensate for the one-sided development based on mostly science and technology. It was a constructive effort through trying to respond to policy and societal needs and through presenting a model for doing so. As Östh Gustafsson notes, however, it was a reactive model, giving the humanities a compensatory rather than generative role. This model has direct parallels in examples discussed earlier in this chapter, including the WASP-HS research program in AI and autonomous systems as well as the notion of a subsidiary model of the humanities (helping other fields and providing transferable skills for future professions) in the US context.[157]

A more resistive response started to develop in the late 1960s and 1970s in Sweden, partly going back to an already established sense of marginalization but gaining ground through several factors, including increasing frustration with higher education policy work, continued discussions of capping humanities degrees, student activism, and the growth of a critical research tradition. Östh Gustafsson argues that this outsider position, which stressed the specificity of the humanities and their threatened position, was increasingly cultivated as an active strategy.[158] It was also an explicit pushback against the more accommodating stance associated with the subsidiary position discussed above. There are parallels in recent pushbacks against "emerging" fields such as digital humanities and urban humanities from established forms of humanities.

Out of this resistive, critical, and inward-focused soil grew an outside critical model in Sweden and the Western world, which has been very influential in shaping humanistic responses to external interest. It connects both to a liberal tradition emphasizing the role of the humanities in educating critical citizens and to an intrinsic-value position according to which the humanities do not need to be legitimized externally.[159] In its extreme form, this model is characterized by a contrary position, disinterest in engaging constructively with society, and a sense of ongoing crisis. This is an ideological and institutional stance. Victoria Höög pointedly argues that the conventional humanities have been unwilling to admit to having a societal role.[160]

Östh Gustafsson's comprehensive and thoughtful study also demonstrates how challenging it can be to move from a critical deconstructive modality to a constructive modality. The main thrust in his monograph lies on the former, but the author does address the role of the humanities moving forward at the very end under the rubric "A new narrative for the humanities."[161] It would seem to make sense, as Östh Gustafsson indicates himself, to draw on his historical analysis and framing when looking ahead, not only in terms of postulating possible future narratives but also to consider how such narratives can be created.[162]

Instead, Östh Gustafsson references the Socratic gadfly or "stinging" (as well as some other historical models) as a frame although he emphasizes that there is not one but many possible narratives in a space that is characterized by constant renegotiation. However, the reader is not really offered a narrative or a set of narratives. Östh Gustafsson does mention the Anthropocene as new historical phase where the human has become more foregrounded and how we need knowledge about humans as well as technology and science, which can give rise to new meanings for the field as well for developments in new directions. However, he does not speculate what such directions might be. The argument does not appear to be actionable.

As we noted in chapter 2, there can be a tendency for well-meaning constructive efforts to stay within or remanifest a conventional framing of the humanities, including a traditional disciplinary configuration. Östh Gustafsson's invocation of the Anthropocene can be taken to represent a traditional separation of human knowledge from knowledge associated with technology and science, and moreover, it does not engage with the substantial critique of this concept, including its strong anthropocentric tendency—a critique that could, in fact, be used as a driving force for reframing the humanities.

Packaging the Humanistic

I suggest that one way of mediating this zone in between criticality, edginess, and application is to more strongly package and instrumentalize the humanities and related work. Packaging in this sense is about putting things together in some coherent way that makes sense to different actors and collaborators. There are many things that can go into such packaging, some of which can be thought of as instruments and tools.

Tools have a very long history, of course, and the idea of instruments or tools comes with conditioning and expectations. Some things are more likely to be called tools than others, Writer Moira Weigel points to how "tool" contemporarily has come to be associated with digital tools and a Silicon Valley–driven notion of such tools as apolitical and somehow existing outside the realm of politics.[163] As chapter 2 demonstrated, research infrastructure speak is full of "tools." Scientific instrumentation has a strong connection to experimentation and science, just like research infrastructure.

A rich literature demonstrates how scientific instruments become part of the complex "packaging" of science. We looked earlier at Bowker's analysis of the Schlumberger method for testing potential oil fields.[164] In their classic study of Robert Boyle's air-pump experiments in the 1660s, history of science scholars Steven Shapin and Simon Schaffer point to the machine (the air pump) as a means of intellectual production as well as an ontological metaphor for Boyle's mechanical philosophy.[165] Frames are made and can be shifted. In an earlier examination of research infrastructure, I argued for the notion of a "humanistiscope" to both attach to and shift the material instrumental framing of infrastructure and encourage alternative projections of what humanities infrastructure can be.[166]

For example, whereas visualizations made by visualization tools are often taken as "ready-made," visualization efforts driven by the humanistic can engage with challenging basic notions, such as the representation of time and space and the reliance on the single screen, and explore interventionist, feminist, and postcolonial strategies in order to test and enact multiple epistemic or political positions, foreground otherwise unrepresented data, facilitate increased interpretative strength, or bring up questions of labor and responsibility.[167] All tools embed inflection and positionality, often implicitly, and this "space" can be recruited to create tools that overtly do different types of critical and constructive work.

What is important is that underlying assumptions and intentions are on the table as much as possible rather than taken for granted. Digital tools can also allow experimentation with different viewpoints, positions, and datasets within the same environment. Data can be superimposed, layered, juxtaposed, reversed, deoptimized, recontextualized, or challenged. Multiplex visualization can move beyond single large screens or immersive environments to arrangements that can afford different types of narratives and enact multiple epistemic positions.

It would seem untenable to imagine a single tool that would suffice for humane capacity building of the sort discussed in this book unless that tool was something very generic, such as a critical or humane perspective, but if so, the tool would likely be too generic, uninteresting, and undecipherable to be useful. Tools in this context arguably need to have some degree of definition, attachment, and constructive reach. Critical, feminist, or multiplex visualization could be such a tool or framework just like infrastructural inversion, staying with the trouble in design, prototyping social forms, posthuman design, glitch, forensic architecture, and perhaps crip technoscience (it if it is not "too big").[168] Some tools could foreseeably also be smaller, such as event formats (e.g., critique sessions), ways of collaborating with students and faculty around themes (such as the Humanities Lab[s] model at Arizona State University), and outcome-driven models such as infrastructural recipes.[169]

Capacity building requires versatility and scale as well as the ability to connect to the situatedness and particulars of each "case." Different tools will be useful for different situations. Such a multiplex approach also avoids the risk of getting locked down by single frameworks. There needs to be a willingness to work across epistemic traditions and methodologies and to add and possibly remove tools over time. Tools must also be decipherable and suggestive to different actors to facilitate conversation and collaborative action.

It has been argued that tools—in this case, strategic tools—are "focal points around which knowing-in-practice arises" rather than being exclusively instrumental and solutionist.[170] This is also why interpretative flexibility is important, which is compatible with regarding tools as boundary objects that exist meaningfully in multiple intersecting worlds.[171] Although such tools need to be flexible and have some kind of generic use, they also need—as noted above—some degree of sharpness and definition, as too much vagueness is likely not to provide a sufficient "grip" or purchase on the perspective at hand.

The two humanistic perspectives discussed at the beginning of this chapter—terraforming and infrastructural inversion—would seem to be good candidates as tools in multi-actor situations seeking to respond to challenges and problems. For example, terraforming as a tool can help us rethink the relation between nature and culture in urban environments without getting too stuck in the existing infrastructure and can raise questions about the way that nonhuman actors have conventionally been handled in urban environments. Infrastructural inversion can be used to open the ethical dependencies of a company's algorithmic systems and pose questions about matters such as inflection, labor, and logic of these systems and the underlying data.

It is not enough for such tools to raise questions, but they also need to lead to a process with some detail, situatedness, and critically informed suggestions moving forward. With a "biased" algorithmic system, such a suggestion could be to change the underlying algorithm and data rather than to attempt to make it unbiased ("fixing" it), potentially making the relevant categories and perspectives be part of the logic instead of just an amendment. Missing information in the data—the structural gaps—could be used as a driving force. New formations could be prototyped and tested on company infrastructures.

Imaginary terraforming in urban planning could help rethink how humans and nonhumans interact and move forward together. For example, how can water—as traditionally represented and regularized—be imagined as a shared and accessible "life blood" instead of a centralized and concealed system for hygiene and water distribution?[172] How can urban farming and local energy logistics change the city as well as challenging standard templates and imaginaries for city projects? And how can we rethink basic urban infrastructure on a system level through infrastructural middleware operating in between legacy infrastructure providers (electricity, water, etc.) and other actors such as municipalities, housing companies, and city dwellers?

Toolboxes

Having many tools rather than one or a few strengthens the sense of instrumentality and level of sharpness. Whereas we can talk about a conceptual tool in an analytical framework in a rather vague way, the invocation of five or six tools forces a different type of understanding and, as noted above, also choice. There can be an open and emergent quality to such collections—essentially infrastructures—as long as they retain their sharpness and avoid becoming something that looks too much like a consulting firm's box of "solutions." This is why the integrity, diversity, and edginess of the tools are important as well as the need to reflect on the tools and their coming together.

Media studies scholar Shannon Mattern points to how not only tools but also the way they are put together comes with values and inflection:

> **A kit isn't just a box of supplies; it's a judiciously chosen collection of tools and materials designed to script a particular process, aimed to serve a particular purpose (or purposes, plural), and to do so with minimal waste and frustration.**[173]

Any conception of toolsets for humane infrastructure building would have to consider the epistemological, ontological, configurational, and values-driven properties of such constructs. There is a risk of ending up with a toolkit that is not much more than a design thinking card deck. Almost any move in an instrumental direction will face some of these challenges. There is, however, also the potential to create something that can serve an important purpose: the bringing together of multiple perspectives, methods, and theories in such a way as to make them decipherable without losing their edge or interpretative power. Such efforts must be self-aware, grounded, and results oriented.

One way of counteracting the streamlining of toolkits is make sure that the included tools are quite different in terms of their "size," process, and epistemology. They are there for a purpose, of course, but their shape, form, or function is not a given. Also, a strategy for avoiding the particularity and familiarity of "toolkit thinking" would be to call the collection a "toolbox" rather than a "toolset" or a "toolkit." The concept of a "toolbox" seems more generic and multipurpose, and it arguably also forces a more concrete image, which can help when instrumentalizing abstract and complex approaches.

Finally, as an expanded note on the "size" of tools, a toolbox can include many different "things"—for example, methods, theories, and perspectives. The tools can be precise or extensive, but we may want to avoid very small things, such as a specific model for sentiment analysis of texts, and very large things, such as a theory, a comprehensive framework, or a worldview. The small things are too small to function as suggestive and decipherable boundary objects, and the large things do not let themselves be easily instrumentalized, negotiated, or integrated with a set of tools. I suggest that some of the big things (e.g., a postcolonial worldview, quantum capitalism, or critical making) address fundamental questions, concerns, and processes that can be part of setting the stage for collaboration rather than being tools (see also chapter 4).

Some Takeaways

What lessons can be learned and inform the discussion of capacity building in chapter 4? One takeaway is that *maintaining the integrity of the humanistic and humane perspectives is central.* What happens in many collaborative situations is that the sharpness and edginess of the humanistic gets watered down or redressed. I argue that the value of the humanistic lies in not compromising these basic positions in ways that remove their strength. Such compromising can be imposed externally (e.g., as part of how the humanistic gets invoked systemically, for instance as "companion researchers") but also from within the humanities. Sometimes, the problem is overly adapting to a perceived need, in a sense "over-instrumentalizing" the humanities.

Humanistic integrity in this sense is not about backing into a corner and defending some golden pure version of the humanities.[174] Rather, it is about engaging, changing, and negotiating the position of the humanities without giving up the idea of the humanistic and humane. Such an approach requires the humanistic and humane to have a place at the table—not be a second-thought flavor to be attached or a ready-made perspective to be delivered—but also to be willing to change and adapt. This leads me to a second takeaway: *Humanistic knowledge is not separate from its potential impact*. Even when humanities knowledge is understood as having a direct influence on the world, it is often modeled as something that gets taken to the world rather than something that gets created and shifted together with others in far-reaching collaborative processes. We need an "attached," intellectually driven, and responsible approach to humanistic work

Such an approach requires an expansive notion of what constitutes humanistic knowledge work, which is a recurring point brought up in the emerging humanities. As I have argued, the humanities have to be involved in designing and building as well as in critiquing and contextualizing. Conditions and infrastructure thus need to change, but for this happen at any scale—and this is another takeaway—*the humanities need to become much better at framing themselves*. Such framing is always strategic to some degree, but it needs to be true, honestly forward looking, and driven by curiosity and the kind of engagement this chapter has explored. Stepping up does not have to mean selling out, being compromised, or upending all what is there but rather taking the opportunity to imagine and frame trajectories, roles, and futures, not only for the humanities but for the academy in a cultural and social context.

The humanities are never going to transform the world by themselves, and not all humanists (broadly speaking) are interested in societal impacts of their work and engaging extensively with actors outside the university. Instrumentalizing the humanities is not about all humanists suddenly changing the way they do their work or about them going out into the world in large groups but rather about encouraging engagement, using humanistic knowledge, and being part of transversal arenas. Curricular development is one useful direction, especially programs that extend across disciplines and connect to the outside world.

The work required will almost always be collaborative and cannot merely be the humanities reaching out with their knowledge, being nice or sparkly, or being attached as an accompanying researcher after the project or initiative has been started. This brings me to a fourth and final takeaway: *Humane and humanistic capacity building must be carried out with others and requires facilitation and curatorship.*

The focus of this chapter has been the humanities and—to some degree—the social sciences. While there is an obvious rationale for this focus for humane and humanistic capacity building, we need to be clear about the fact that such capacity building is not about creating humanities think tanks. Instead of encapsulating and rarefying such knowledge, we need to activate, shape, and engage it in broad, thick, and respectful collaboration.

In any case, it does not seem realistic to imagine a world full of humanities-based construction firms, programming teams, and hospitals. The humanities will likely not provide any comprehensive solutions by themselves, but they can be central to coming up with responses—different responses—together with others inside and outside the academy, including engineers, programmers, designers, activists, and government representatives. This requires us to see collaboration with other disciplines and other actors not as something we add to the humanities but rather as a core value that we enact through the work we do. The humanities cannot act by themselves. They are part of the university not just organizationally but also conceptually, part of a university that extends far beyond itself.

4

Designing Humane Infrastructures

In this concluding chapter, I turn to the question of how to build capacity to mobilize humanistic and humane knowledge to respond collaboratively to challenges and problems. Essentially, I believe that this question is relevant to anyone interested in making our world a better place, which is why it is worth taking on, despite its illusive and immense nature.

The invocation of "capacity" is a way of emphasizing the need for sustainable, comprehensive, and recognizable efforts beyond think tanks, design methods, and singular centers. Such capacity must have certain qualities. For example, it must be reasonably generic; change with stakeholders, communities, and collaborators; and be as open, networked, and distributed as possible. It must also be self-critical and sensitive to the tension and critical inflection of its own conditioning. Such capacity can ideally be seen as a humane infrastructure itself as well as an arena for engaging with the reshaping of infrastructure broadly speaking.

The concept of "capacity" is useful, as it is broader and more responsive than "infrastructure," although there is considerable overlap. Just like infrastructure, however, capacity is a laden term. It is often associated with management, fundraising, and a tendency toward homogeneity.[1] Unlike infrastructure, however, capacity building is more strongly associated with mobilizing resources to respond to imminent needs. One reason for this dynamic quality is probably that "capacity" is less bound to material framings, policy systems, and timelines, which potentially makes the concept more amenable to shaping or reframing.

Infrastructure is lived experience, and building capacity is about individual circumstances, personal commitment, and communal engagement. Because of the structural and institutional nature of things such as capacity building, which will be explored in the following under rubrics such as "conditions," "frames," and "arenas," I feel that it is particularly important to let the account touch ground. My approach is inspired by anthropologist Kathleen Stewart and her invocation of a contact zone of ordinary affects as a means not to lose track of the "reeling present" with its heterogeneity and non-coherence for approaches that merely "slide over" the surface of the ordinary to get to the big structures and underlying causes.[2] It is important, for example, to give more than a little space to multivocal and situated conversations, creative and experimental practice, and diverse case studies that do not necessarily fit neatly into the structure—indeed, they can interrogate and challenge those very structures.

Interlude: Shifting Sociotechnological Systems

Carrying out work that is simultaneously critical and constructive requires extensive groundwork not to get frozen in distanced theorization, templated production, or institutional conservatism. Such groundwork—which feeds into capacity building—takes time, negotiation, and attention to intellectual and material details. I will draw on a conversation among a group of scholars and practitioners to shed some light on this groundwork. This conversation—on changing sociotechnological systems—was carried out over the summer of 2021 with two main conversationalists as well as three discussants and the current author as the curator.[3] The main conversationalists were Marie Louise Juul Søndergaard (MLJ), based at the Oslo School of Architecture and Design, and Matt Ratto (MR), based at the School of Information Studies at the University of Toronto.[4]

I contacted Juul and Ratto because I had recently heard them each talk about a current project of theirs that seemed to engage with making in a carefully critical and generously reflexive way. These projects were not one-offs but rather part of longer trajectories. Ratto comes from ethnomethodology, science and technology studies, and design, whereas Juul has a background in digital design, feminist technoscience, and interaction design.

Marie Louise Juul's project was a collaboration with researchers in micro and nano systems at KTH Royal Institute of Technology. She worked with an expert in microfluidic systems and point-of-care diagnostics who was developing a home test for women for urinary tract infections. The collaboration intended to make the urinary sampling and testing process easier to carry out for women in their homes. Juul, with a strong interest in women's health, femtech, and menstrual and sexual technologies, had previously done speculative design projects about menstrual blood and urine analysis and was very interested in the scientific and medical side of the home test project. Her vantage point was that of "troubling design," drawing on Donna Haraway's work as well as aligning with Queer HCI and feminist design.[5]

Juul stresses the role of the collaboration and how much she learned from the collaborators, not least in terms of how tools shape and enact knowledge. She did not come to the project to change its worldview and design overnight, and conversely, her collaborators chose to work with her even though she did not come with the conventional design and user perspective they may first have expected:

> **Marie Louise Juul:** What also became apparent for my collaborators was that my critical and feminist-informed design perspective was not about making the tools easier to use or more beautiful in the eyes of women (as if I could ever represent such a plurality of experiences). While this might be the dominant idea of what design is and does, I believe I managed, through a number of meetings, workshops, and design provocations (such as sketches, scenarios, and prototypes), to open up for reflections on how this tool would act also as a sociotechnical system that positions women as patients, bodies as under medical scrutiny, and the home as a hospital—and eventually to trouble this narrative and imagine how it could be designed otherwise. So for me, this collaboration was never meant to result in finished products that could be marketed and sold, although that might be the case for the wider research project that I intervened in, but rather to better understand from the inside how medical technologies are developed and what role critical thought and design practice can have in this.

There is a noninstrumental sentiment here that is central. It is not in contradiction to making useful tools and systems but rather about resisting the conventional instrumental framing of sociotechnological tools and infrastructures—including the presumptions embedded in that instrumentality—and potentially changing what counts as useful or as a good tool. Juul critiques current models of femtech as reproducing dominant values of technology, including datafication, centralized power, personalization, and disembodiment. The operating frame of these models is health issues as problems to be solved by technology.

Shifting such models or frames requires the participants and stakeholders to be willing to negotiate and change their positions. Juul explains how she had to balance the normative ideas of designing—what is typically expected—with moving beyond those norms in the project and that she had to question implicit assumptions as well as reimagining ways of designing not only the technology but also the relations and contexts in which it would be used. There is moderation, respect, and carefully opening up in Juul's work *as well as* a radical sentiment. This groundwork is helped by concretely bringing epistemic tools and practices to the table, which, in Juul's case, included workshops, sketches, scenarios, and prototypes. In the conversation, Matt Ratto points to how prototyping is a connective practice ("acts of translation" to bridge epistemic gaps between critical and constructive work) underutilized by many critical scholars, partly because there is often not adequate material and intellectual infrastructures.

As Ratto notes, such multistakeholder work by necessity involves being explicitly complicit in the results of the shared work and being responsive and responsible for what is created in the collaborative process. Just like Juul, he stresses the importance of reflexivity and self-awareness, not only in relation to stakeholder positions or humanists engaging with design but also as part of technological work more broadly and within the tools themselves. Juul seems to go further than Ratto, however, when describing tools as change-making instruments that can invite a certain direction.

Ratto has extensive experience of such tensions and the domain generally through having developed "critical making" as a framework over more than a decade.[6] The original impetuous for his work in this area was a concern with the linguistic and discursive basis of materially engaged critical theory as well as an interest in exploring, analyzing, and teaching the often downplayed critical and conceptual work embedded in making. Over the years, Ratto has moved from a process-focused interest, where the result was not really that important (even demonstratively unimportant), to a more a direct engagement with making long-lasting change in the world—a direction exemplified by his group's work on 3D-printed protheses. The project he had told me about a few weeks earlier was medical in nature like Juul's but placed somewhere in between the academic world, a hospital, and a form of counseling:

> **Matt Ratto:** I have a sort of similar story, although I have to say I wasn't invited to participate as much as I found a way in. . . . I've been interested in the general area of AI and behavioral intervention and was looking for an interesting project. A colleague of mine in computer engineering has been working for the last few years with a clinical scientist and professor of medicine who runs an addiction program at a Toronto hospital. They have been building versions of a conversational agent that uses natural language processing systems to provide automated counseling to help people quit smoking. This project seemed particularly interesting due to the way the bot is intended to replicate a counseling

form called "motivational interviewing"—a client-centered, empathy-based counseling process. . . . We've been participating in the life of the project for a few months now and have provided some help in thinking about how trust is considered in the design of conversational bots. My sense is that our biggest value is in providing a deconstructive lens to the project and reopening up previously under-considered opportunities. For example, recently, we have been thinking about the criteria of "humanness" in AI and conversational bots and leveraging feminist posthuman starting points as an alternative to typical design choices. For me, the key to this work is not just in denaturalizing things such as "humanness" but also in utilizing this to provide operationalizable insights that help resolve issues the broader team is facing.

The reopening of earlier opportunities that Ratto references is central to his approach and something that directly relates to the groundwork done by researchers, designers, and stakeholders. Instead of aligning change and innovation with displaced visions mainly driven by technology, Ratto argues in favor of grounded and thick versions of hope—"ordinary hope"—building on his collaboration with Steven Jackson.[7] There is a parallel here to Juul's description of how her contribution to the collaborative project "thickened the context" and how she talks about avoiding generalizable and too-big theory.

Reopening past choices can be a powerful strategy for shifting a design process and affecting the texture or logic of a project. For instance, considering more-than-humanness in the counseling project has led to different ways of conceptualizing and building the conversational agent. One example is making the agent less humanlike and more "unrealistic" to improve the quality of the counseling.[8] Just like with Juul's project, Ratto's work challenges what a tool is or can be as well as some of the basic presumptions of the problem complex in question—in fact doing reframing or framing work.

Juul's and Ratto's approach requires being willing to listen, engage, and change as well as holding on to things. Frames can be pushed, tweaked, and shifted as a constructive move but do not always need to be shattered, at least not all at once. This is an important point made by one of the commentators, Sara Hendren at Olin College of Engineering:

Sara Hendren: I thought you both treated the necessity of theoretical and pragmatic work so beautifully, and I want to underscore something MR said in this vein: that lots of folks can identify and critique the excessively instrumentalist logic of engineering, but fewer people admit the failings of, as you say, "didactic objectives which can also be a way of short-changing deeper critical and reflexive practice." I think we all share a dissatisfaction with the didactic mode, especially when objects produce a purely "deconstructionist" narrative! I think often of Svetlana Boym's distinguishing between estrangement in its classic Brechtian sense—here is this aesthetic object in which all the viewer's operative norms are shattered, the end—what she called "estrangement from the world" as opposed to "estrangement for the world."[9] This latter sense is the kind of estrangement, seeing things differently, that still returns us to the broken world as it is and to each other—another example of moving beyond the purely critical and toward possible alternatives, as MLJ mentioned.

As Hendren points out, we need to find ways to engage critically that allow us to move toward possible alternatives. It sometimes seems that the critical edge of the humanities is programmed to critique and take apart certain aspects of sociotechnological systems, whereas it often has difficulties looking at itself critically or moving close enough to engage in constructive reimagination and remaking of those systems. As we saw with the Yale AI response discussed in chapter 3, there is a risk for the humanities to lock off their potential impacts.

Hendren is willing to take on a role of a reformist incrementalist thinker, which may make her "Not Satisfyingly Critical Enough"—a role that she attributes partly to her intellectual disposition and also to raising a child with Down syndrome who needs "new narratives, new political structures, yes—and also hardware and software and services in the better-mousetrap form."

Change takes place across many registers, including narratives, politics, and design, which can appear to constitute different domains but are also intertwined and codependent. Ratto argues that designed objects embed "idea stuff" (such as values, norms, intensions, and why not narratives). One of the other commentators, Nicole Anand, political economist, governance specialist, and participatory designer, maintains that there is not necessarily a difference in scale between governance processes (such as budgeting, policymaking, programming, service delivery) and designing products. She argues that we need to work across these registers as well as different types of structures.

Nicole Anand: Back to the rules, because we all play by different rules. As such, we can believe that the "constructive work" to positively influence societal

> change is both to design within the formal structures of institutions and policies and to design with the "obvious" of some that is also not formalized, widely valued, or well understood. There is value to be added to the latter in all forms and spaces of design that go well beyond the governance sphere. . . . I like to think of these spaces and practices that influence social change as unseeable to many but unforgettable to some. In other words, in the BIPOC and Global South communities that I tend to spend time within, we often say what is obvious to us is not obvious to others.

Anand reminds us that spaces of coming together and change are not singular or all the same but rather multivalent and situated in different ways. While HUMlab at Umeå University was an open meeting place around culture, the humanities, and technology, it was inflected by the location, the privilege of being at a Swedish university, the demographics of the university, institutional and art participants (rather than other parts of the city, which were more demographically and socioeconomically mixed), and the way it was set up conceptually and materially. This is one reason why self-awareness must be a condition for humanistic and humane capacity building.

The participants in the conversation all positioned themselves in between disciplines, epistemic traditions, domains, practices, and structures. Ratto, for example, stresses the need to increase the work in the middle to carry out "experiments and practices of transduction and translation" to bridge different forms of technical work, ranging from the instrumental and reductive to the critical, imaginative, and expansive. This liminal position relates directly to the notion of arenas for transversal knowledge production as explored later in this chapter. Juul emphasizes the importance of working on different levels—including infrastructure, people, and norms—and establishing and maintaining infrastructures, tools, and spaces that support inclusion and which are open to uncertainty and emotions as well as healthy and generous collaboration. Such epistemic and value-driven commitments form essential conditioning for capacity building.

Hendren uses "undivided house" as a metaphor for her lab at Olin College to emphasize that it is a "big capacious canopy" for working on disability/assistive tech with a joyful durational set of practices. These lead to partnerships, projects, and results such as critical design as well as on-the-ground participatory design models. While this is an institutional setup, it is different from many other programmatic efforts. It has a strong directionality as Hendren seeks to "embody the ideals and politics of disability studies *in the lab*" (original emphasis). Furthermore, there is an emphasis on collaboration across epistemic and other traditions, not as an inclination or part of a configurational phase but rather as a persistent driving force. The in-between aspect of the operation continues to be central, and the "how" of liminality is an ongoing challenge.

Arizona State University mathematician and media artist Sha Xin Wei, another commentator, uses the notion of "atelier" or "atelier-lab" to describe a place where disciplines meet thickly and participants reorient their approach in what Sha calls a transversal knowledge machine.[10] One framework or tool used is prototyping social forms, which allows experts and stakeholders to experiment with, vary, and scale social forms computationally and poetically. Gesturing to an artistic space opens up opportunities in framing the enterprise radically and constructively at the same time, including an explicit noninstrumental sentiment and a commitment to creative practice as well as non-digital modes of articulation and a broad range of sensory modalities. Individual projects should not have to be motivated in relation to disciplinary, funding agency, and other expectations, as the overall platform—the atelier—can respond to such rationalistic impulses. This absorptive or protective function has been observed in other experimental institutional setups such as ACTlab at University of Texas and HUMlab at Umeå University.[11]

Another set of metaphors and models were used in the discussion to describe not the environment (atelier, undivided house, etc.) so much as the intermediate actors that seem central to many in-between operations and endeavors—actors that have their own "content" focus as well as a liminal ethos and interest in taking on an active curatorial or faciliatory role. Juul references the shape-shifter as a model—someone who can change form and identity "as conceptualized in mythology, folklore and speculative fiction (for instance the witch)." Different roles and identities come with different sets of expectations, license to do things, and access to contexts, tools, machines, and materials. Ratto invokes the bumbling detective, always listening, ready to get his hands dirty, redirecting attention, generating insight when you least expect it (while hedging for the potential "hero's journey" part of this model). Ratto also adds the following comment:

Matt Ratto: Yes, the "seeing as" is so important. Not just turned inward as a way to understand our own positions and perspectives (shape-shifter, baroque personality, etc.) but also learning to understand and inhabit the "seeing as" of others, in particular (I think) in the ways they (and we) have been "disciplined" by our epistemic and ontological contexts. Shape-shifter, yes, but also epistemic boundary crossers and translators.

Understanding our own positionality as well as that of others—in fact, juxtaposing, overlaying, and realigning them—is a difficult individual and collaborative task, perhaps even impossible, but one that can be helped by intermediators such as Juul and Ratto (and Anand, Hendren, Sha, and many others) and by networks and arenas that support and aggregate such work and knowledge, building long-term capacity.

Creating Conditions

Knowledge production is conditioned through our backgrounds, fields, domains, tools, regulations, and culture whether at universities, companies, or NGOs. A central idea behind this book is that we need places, nodes, and arenas (capacity) driven by the humane and humanistic, which can help us approach challenges and problems collaboratively. This idea, even in this barebone version, comes with a set of conditions, and any further detailing will naturally add more conditions and conditioning to the model. Approaching issues collaboratively ("collaboration") is thus a condition, which I will use as an example below. Conditions are never made from scratch, but they are also not a given.

Using the term "conditions" provides a way of talking about the texture of capacity building and capacity—which typically incorporates many conditions and complex conditioning—while maintaining a critical and constructive impetus. They allow both a critical reading (examining the conditions and the conditioning) and a potential constructive engagement (conditioning, creating conditions), especially if conditions are taken to be under negotiation rather than set. Finally, in its active form—to condition—the term has some potentially negative connotations (controlling, constraining) that can remind us of what is at stake when imagining and building things.

Collaboration

Conditions need to be transparent and on the table. Collaboration, for example, is not in any way a given. As we saw in the case study of humanities centers and some similar institutions in chapter 2, the idea of collaboration can be limited to certain modalities, traditions, and spaces. The compulsory lunches and seminars at an advanced institute alongside the individual offices at a closed-off facility make for certain types of collaboration and not others. Similarly, some public humanities work tends to construct collaboration as the humanities giving knowledge to the community rather than learning from them (chapter 2), and emerging humanities initiatives often emphasize collaboration at the expense of conventional modes of knowledge production (chapter 3). For extensive multipart collaboration to be realistic, there has to be acceptance—and ideally support for—different epistemic traditions and commitments as well as different backgrounds and understandings. This is part of the conditioning of humane capacity building.

Multistakeholder collaboration is about managing such differences, including ones to do with resources (time, money). In an illuminating discussion of multistakeholderism, Nicole Anand argues that well-resourced sectors and experts will have to share power.[12] How such collaborations and concerted efforts are set up is obviously important, and Anand argues that we need to move beyond established relationships and our comfort zone. She devises key strategies—essentially conditioning—for carrying out such work, including embracing being a novice, forming uncomfortable relationships, channeling skepticism productively toward collaborative and critical experimentation, and engaging different lenses to support practice also when this necessitates an "uncomfortable set of deep conversations."[13]

Engaging with people and organizations that have values, belief systems, and ways of doing work removed from one's own is challenging and can certainly be uncomfortable at times (but also potentially rewarding). It takes effort and a willingness to take risks to collaborate with partners who are different from you and where there may be limited or unclear immediate gains and few set pathways. Anand asks us to set up alternative forms of conditions. Such work can be described as curatorial and cultural and is central to the current chapter (see the section on arenas for transversal knowledge production below). How do we create conditions for constructive exchange and long-term change across epistemic traditions, perspectives, and interests?

Conditioning operates on several levels. Staying with collaboration as a topic, many universities will have central administrative or service functions that handle community collaboration, development work, and "collaboration" with donors, imposing certain conditioning on collaboration, including gatekeeping and prioritization of certain fields (such as science). Collaboration is also conditioned through how academic work is organized. For example, there is often limited accommodation for working across departments and schools within the organization. A program operating across fields may find it difficult to cost share faculty (because of departmental and disciplinary logic), and access to infrastructure is generally made considerably more difficult if it is based at another school or faculty (different organizations and deans, ownership, etc.).

Scale

Capacity building of the sort discussed in this chapter and book would necessarily include a range of conditions. Another example, aside from collaboration, is that this capacity needs to be reasonably generic rather than restricted to specific challenges or fields to be as cohesive, flexible, and functional as possible. There is tension here between the local, the situated, and the large scale and generic. We observed this tension in relation to the framing of infrastructure (chapter 1), which tends to privilege the large scale, generic, big, and technological at the expense of the local, situated, small, and human. The humanities, on the other hand, tend to emphasize the local, situated, and embedded and be skeptical about generalizing, aggregating, and scaling up. At the same time, humane capacity arguably needs to extend across this range—from the particular and lived to the aggregated—for critical depth, explanatory power, and instrumental potential.

Urban scholars Hillary Angelo and Kian Goh engage with the issue of scale and range in relation to difference and abstraction in planetary (very large scale) urbanization.[14] They introduce themselves as queer feminist women scholars engaged with responding to "difference" as a main critique of planetary urbanization. They find three types of critique—empirical, epistemological, and theoretical critique—all of which can be related to a conflict between abstraction and difference. Empirical critique comes from postcolonial, feminist, and relational approaches to urban studies, which argue that difference is obscured by planetary urbanization. Epistemological critique comes from the position that planetary urbanization extends a privileged (white, largely male, Euro-American) viewpoint. Theoretical critique, despite representing different positions, is said to come together in emphasizing and challenging global systems of domination, including capitalism.

Angelo and Goh argue that we need to explore the ways local politics extends to systemic and structural conditions as well as complicity and alignment across scales in order to shed light on important intersections between the large and small scale that can give us analytical purchase as well as providing leverage for contesting unjust futures.[15]

> **Rather than bringing a charge of "difference against abstraction," we argue that research informed by planetary urbanization can and should continue to traverse spatial scales and analytical levels—not only between the concrete and abstract at the large scale, but also at the small scale and everyday level.[16]**

Bridging scales then becomes not so much a problem as a necessary strategy to approach systemic challenges and structures across scales: a multi-scalar approach. Such transversal work needs to be carried out carefully without losing track of the intersections or flattening the multiple scales at play. An important question is how to move between scales responsibly—for example, when reconciling local environmental data in their situated, historical, and epistemological context with global and actionable narratives of climate change. We also need to be attentive to how concepts such as "scale" and "scaling up" are embedded in models of technological and financial growth and resource management.[17] Humane capacity of the sort discussed here not only needs to be sensitive to such questions but also must be able to sensibly support the action required, itself multi-scalar.

Complicity

The question of complicity raised by Angelo and Goh (and others) has broad consequences for capacity building. Constructive engagement in collaborative processes with external stakeholders requires scholars to be overtly complicit, in relation to both assuming responsibility for the systems in question and responsibility for making such systems better.[18] We are already complicit in systems that are part of our lifeworld, including societal and academic infrastructures, but engaging constructively with them requires us to acknowledge this complicity as a condition rather than operating at an analytical distance or declaring other actors at large (such as universities or corporations) to be unredeemable offenders ("evil"). This can be a challenging proposition, given that critique has often been used to "disavow our implication" in academic (and other) infrastructures, enabling a sense of autonomy, but also because the university is not steeped in ameliorative ideology or potential for everyone but for some rather in continuous coloniality, counterinsurgency, and dispossession, which naturally limits the possible constructive incentive and range.[19] There is a difference, perhaps, between being "implicated" and "complicit" in that the latter seems to presume a larger degree of embedment and responsibility. A third term, "detached," can be used to emphasize an active embracement of being disengaged. As cultural anthropologist Alisse Waterston argues convincingly, detachment is a position:

> **Engaged scholarship reflects understanding that "detachment" constitutes a form of action. To pretend otherwise is to ignore the fact that distance and distancing are not at all neutral but have consequences, a form of complicity with those policies, practices and ideologies that harm those who are most vulnerable and that which is most vulnerable.[20]**

Conversely, not being detached—being "attached"—is also a form of action. Indeed, it needs to be an action, and the shape of that action is crucial for productively combining critical and constructive modes of engagement in humane capacity building. While different types of engaged scholarship and methods such as design-driven prototyping and storytelling can be crucial, they can also appear detached in the sense that they do not directly engage with the challenges at hand.

Attachment also means engaging with stakeholders in their environments, whether local communities or corporate entities, as well as university partners. There can be considerable value to finding ways of working with partners in their environments—for example, sharing, changing, or building infrastructure in situ instead of building new shiny infrastructures on campus. Such collaboration—while often time consuming and challenging—can forge strong links, mutual dialogue, and understanding and lead to more direct results.

There is no single set of conditions that will do the job, just as there is no one solution or approach to complex problems. What is most important is that we are critical and reflexive toward existing conditioning as well as willing to suggest and implement conditions to build capacity. These conditions tend to be complex and numerous, forming clusters of conditions. Such structures of conditions approach frames.

Shifting Frames

Capacity building is partly about framing. Capacity is complex and needs to be put together and presented cohesively. Frames also need to be critiqued and changed. David Theo Goldberg states that critical thinking and critical theory typically assume the frame of what is given and conventional and that these traditions therefore think "*within* the faded frame" (original emphasis).[21] He then goes on to argue that we need to take on "the social and thought frames themselves, the structural conditions constituting the conditions of possibility, the constituting outsides which order established and inherited ways of being, thinking, and doing."[22] Building humane and humanistic capacity requires challenging and changing such frames.

Following Goldberg, the notion of frames is here used to describe the structural conditions that can be associated with infrastructures, whether ideational, material, or social.[23] The challenging, shifting, and changing of such frames require constructive modes of engagement, some of which were discussed in chapter 3, as well as capacity to bring things together, which, in turn, must build on an ongoing negotiation of the humanistic as well as other frames and positions.

Reframing Work

Many frames not only are resistant to change but also appear to be almost unchangeable or unbreakable. Conditions are embedded in the practice, institution, or platform, both encoding and reinforcing belief systems and worldviews. This embedding lends strength and protection for frames, and as Goldberg argues, we tend to stay within the frames, even when we resist or study them. For example, critical studies of infrastructure can sometimes seem to take the outer frame of infrastructure for granted, including the essential usefulness, systemic nature, and large size of infrastructure.

One example of resisting the framing of infrastructure as essentially useful and benevolent is offered by the notion of "toxic infrastructure." Clearly, this type of infrastructure does not fit the infrastructural template, just like "invasive infrastructures" does not.[24] It is simply difficult or perhaps even impossible to associate toxification with usefulness, progress, and monumental achievements, and the slow violence carried out by toxic infrastructures cannot be easily repaired or explained away. Indeed, in her analysis of Jennifer Scappettone's poetry relating to the Syosset Landfill site, scholar and poet Kate Lewis Hood argues that Scappettone, together with her collaborators and interlocutors, "collectively compose a salvage poetics, breaking up the boundedness of 'bodies' and 'toxins' acting upon one another."[25] Infrastructure cannot be disassociated from its effects. How can we repair the bodies and communities affected by toxification?

This type of unboundedness is also exemplified in experimental projects such as architect researcher Nerea Calvillo's infrastructure *Yellow Dust*, where mist was used to make air and pollution visible.[26] Calvillo attempts to shift the focus from a cleaning approach or frame (exemplified by geoengineering projects) to a nonpolluting situation. This model can be contrasted with that of Natalie Jeremijenko, whose focus on improving environmental health is not compatible with (merely) reducing emissions or going for zero-emission strategies.[27] She argues for systems and infrastructures that make things better.

Jeremijenko challenges not only the frame of reducing or getting rid of emissions but also the systems that produce templated infrastructure. In an interview article, she argues that the way public infrastructure is designed is a socialized risk, a risk we all bear, which is why it is so important to consider how infrastructure projects are processed and implemented.[28] She discusses how urban infrastructures get framed as large, nonexperimental, nonlocal, costly, and privatized:

> **How they get designed is through large city contracts awarded to master planners, and the most invisible are the huge contracts to large corporate engineering firms that have a stranglehold on the public purse. They have no incentive whatsoever to experiment. In contrast, I think it's important to talk about small-scale material practices that really do enlist diverse opinions.[29]**

Jeremijenko highlights the local perspective, not least through enacting a series of concrete projects, while also emphasizing the role of experimental practice and imagination in rethinking infrastructural systems at large.

One way she makes the connection between these levels is through using "environmental health" as a frame, operationalized through the Environmental Health Clinic (or the xClinic):

> **Yes. That's probably the biggest frame. . . . Although now I've been putting things more into the Museum of Natural Futures. The Environmental Health Clinic is really a framework. Everything I do might be convivial, or playful or seem silly, but it all measurably improves human and environmental health. So, I think that that's the only proxy, the only metric and the best proxy for the common good is health . . . no one's really anti-health. And so, we have a common ground, if you will, to build things.[30]**

Jeremijenko thus uses environmental health as a frame to connect local experiments to a larger goal and to constructively overcome what she sees as a crisis of agency. This frame includes a number of elements, including environmental health, the associated clinic, the X-teams, "impatients," treatments, and the various projects or cases. Impatients are called so because they are too impatient to wait for change in legislation or health systems. These are collaborative intraspecies tasks. Jeremijenko argues that we can all change our indoor and outdoor air quality, for example, and modular infrastructures such as "agbags"—a vertical urban farming system—has been designed to support biodiversity and improved air quality in urban environments (figure 4.1). The frame gets strength and persuasiveness from the rich and imaginative composition as well as from the experimental practice.

It also becomes clear how Jeremijenko actively engages with complementary frames over time, including the museum of natural futures mentioned in the quote above, which is described as a place for hosting delightful public living infrastructures, including data, distribution, water, energy, waste, and food systems to expose and introduce designs that are said to transform the existing local systems and improve human health.[31] This frame, just like others Jeremijenko uses—mutualistic design, a crisis of agency, the space race of the twenty-first century, and so on—tend to overlap or align, at least to some degree. A key reason for this compatibility is that there is a cohesive set of core ideas, including "cohabitation" and "improving health," a persistent sense of wonder, and a reasonably stable vision despite the multiplicity of projects, engagements, and challenges addressed.

Figure 4.1
Agbags (a system for modular vertical urban farming), part of Natalie Jeremijenko's experimental practice. Photo credit: Natalie Jeremijenko.

The Smallest Acts

At the same time, presenting, enacting, and getting support for such alternative work is challenging, especially in a long-term perspective. There is almost by definition a lack of support and resources for experimental operations of this sort, and many initiatives depend on a few key people. More generally, taking on frames, systems, and narratives in a radical manner can seem daunting, especially since they often appear monolithic, impenetrable, and unresponsive.

In the context of infrastructures—often seen or presented as near permanent and nonnegotiable (How do you reason with an interstate system or an unjust civic structure?)—it is particularly important to emphasize the quotidian nature at the heart of such change. Angela Davis talks about how the struggle for racial justness over time made structures permeable that previously had been—and been seen as—impermeable.[32] Small-scale change and routines—that can be repeated and habituated—are central to many activist approaches.[33] In this also lies the need to enact humane values and principles through how we act in the world. Well-meaning and humane institutional and infrastructural frames or visions cannot just be discursive but must be part of how we live those institutions and infrastructures day to day:

> **It is the difference between what we say we want the world to look like and what we *actually* carry out in our smallest acts. Carrying, how we carry ourselves in our relationships and how we carry each other, is the real place of transformation.[34]**

Digital humanist Marisa Parham's invocation of our smallest acts resonates with Jeremijenko's emphasis on small-scale material practices. Change is processual and has to be made in the present with an eye on the future rather than being driven by displaced "shiny" visions, as is often the case with technology and infrastructure work. There is something about such displaced—often both unrealistic and unwanted—visions combined with the heaviness of infrastructure that can, in fact, prevent change and hope. For example, it is doubtful whether "smart cities" as an idea and vision really has helped urban development (Again, what would a humane city be like?).

Technologically driven visions can also be influenced by the idea of existential risks as part of "longtermism," where present-day "real" problems get backgrounded. It has been argued that such approaches—said to be driven by a tech elite—focus on the "flourishing of imaginary future people rather than the well-being of the masses right here, right now."[35] In a powerful response to a public letter to pause "giant AI experiments," Gebru, Bender, McMillan-Major, and Mitchell maintain that it is more important to address current exploitative practices and a growing concentration of power than to do "fear-mongering."[36] They are right, but this does not mean that we should not be structural and visionary as well as grounded and situated. Challenging and shifting infrastructural frames cannot just be horizontal work, as the templates are also vertical, just like much values-driven work.[37] The constructive potential associated with infrastructure and technology can help us imagine good and caring futures and shape the conditions of possibility referenced by Goldberg, even if there is a risk of displacement.

Urban Infrastructure as Memorial

When Jeanelle Austin, activist, community leader, and public intellectual, was made aware of the murder of George Perry Floyd a couple of blocks from her childhood home in Minneapolis, she started to text family, colleagues, and friends from where she was in Texas, identifying resources required to "protest well, as a peaceful protester."[38] Austin contributed to building capacity in this manner and soon became engaged in tending to the George Floyd memorial, arguably an infrastructure.

The memorial pays tribute to the memory and life of George Floyd. Floyd was killed by a police officer outside the Cup Foods store at the intersection of 38th Street and Chicago Avenue in South Minneapolis in the presence of three other police officers and a small crowd, including a young woman who filmed the murder on her phone. The memorial was created spontaneously after the murder and grew quickly—a center point at a time of massive turmoil. It extends over an urban space and a major throughfare. Car traffic ended up being closed for more than a year.[39] A projected metro station at the intersection was not built.[40]

The community took over the space and made it into an "autonomous zone" and, in Austin's words, a space for public grief and holistic restorative justice. Guidelines and practices were established around the memorial—for example, treating everything left as offerings not to be thrown away and regarding people as more sacred than the memorial itself.[41] The contributions are not cataloged, but one assessment is that there are about two thousand stored offerings alone.[42] Installations were added around the space—for instance, "Say Their Names Cemetery" by Anna Barber and Connor Wright, where 132 lives are memorialized through boards cut out in the shape of tombstones.[43]

The George Floyd memorial can be seen as infrastructural, part of what Charmaine Chua has called "alternative infrastructures of care and repair."[44] It is an organized space for public grief and restorative justice materialized through the offerings, installations, and community engagement in an immensely charged location in a city with deep racial tension in a country whose history is one of profound racial injustice and racial violence. It also became a platform or capacity for negotiating with the city and for the city to engage with the community, as when Austin and other community members solicited definitions of "justice" to produce a resolution with twenty-four demands on the city for concrete actions of justice.[45] The memorial is not fixed, unitary, or tidy and neither are the community's ideas about justice.

The material expanse of the memorial (and, more broadly, the intersection and what is now called "George Perry Floyd Square" between East 37th Street and East 39th Street) gets linked to an emerging ideational and communal underpinning in a process of continuous reframing and institution building (as expressed by lead caretaker Jeanelle Austin):

> **My job requires facilitating an imagination, not just around how do we preserve and hold the story for generations to come, but also, how we build an institution within our social imagination where we practice justice within its institutional DNA.[46]**

There is alignment here with Jeremijenko's appeal to imagination, and both she and Austin are committed to the material ramifications of that imaginary work. In the case of the George Floyd memorial, the weight of the murder and the protests certainly forced the city to rethink their investment in racial justice and communal engagement, but the fact that there was a piece of infrastructurally heavy urban space at stake was also significant. The space both put pressure on the city and offered a platform for channeling actions and initiatives. Eventually, the project "38th & Chicago Re-Envisioned" was initiated by the Minneapolis Public Works department:

> **In the next few years, Public Works staff plan to reconstruct the intersection of 38th & Chicago. In 2022, we will begin a process to engage with the community. Through this process, we intend to redesign the area in a way that reflects community needs. This is a significant opportunity for us all to define the future of 38th & Chicago. . . . In 2020, George Perry Floyd Jr. was murdered at 38th & Chicago, forever changing the intersection and surrounding community. This area is known as George Floyd Square (GFS), and it has evolved beyond its original roadway design.[47]**

This description is an example of what may be called "hesitant reframing," somewhat akin to when the framing of research infrastructure increasingly had to incorporate the humanities (chapter 2). In this case, Public Works importantly not only acknowledges the murder of George Floyd and its impact and importance but also holds on to their sense of infrastructure ("original roadway design") and ownership of that infrastructure ("Public

Works staff plan to reconstruct . . .") and suggests a radical but local change rather than a change with a transformative impact on the whole city and beyond ("forever changing the intersection and surrounding community").

The ensuing process orchestrated by the city mixes different types of community engagement, themes, and methods—including asking participants to create a street through laying out standard street puzzle pieces (including car lanes, outdoor dining areas, and different-sized gathering spaces)—but however stumbling some of these design moves seem and however much rightful community resentment can be found, there is a collaborative reframing in process.[48] There have been some policy changes and limited investments, and while these are not nearly enough or community oriented enough, the city now seemingly has a commitment to "balance traditional asset management needs with the intersectionality of justice, healing, placemaking and culture."[49] There is no way of rightfully assessing the situation from the outside, but it does seem that the frame has shifted—that there is dialogue (albeit partly controlled by design methods and troubled by deep and rightful distrust) and hope associated with changing the infrastructure, broadly speaking.

Such work is difficult for many reasons, including the necessity to engage with partners that are seen by some as irredeemable offenders, communities whose ideas about urban space do not fit the infrastructural template, or structures that almost certainly will remain structurally unjust, even if they are changed for the better.

Porousness as Opportunity

The framing of public works and infrastructure in Minneapolis was thus shifted through the murder of George Floyd and the work of Jeanelle Austin and many others, including activists, community members, and city officials. Infrastructures—if challenged—can be more malleable than they might seem at first sight.

Critical theorist Lauren Berlant sees opportunity in the porousness and stretchiness of infrastructure, particularly in a time of transition, in the long meanwhile of crises. They argue that if a glitch has surfaced conditions of disrupted resource distribution, jurisdiction, and circulation, a disruption in rules and norms cannot be equated with the defeat or absence of structure as such.[50] Infrastructure becomes a way of moving away from our tendency to see structure as solid, particularly at a distance. Infrastructure can then be seen as "productive, durationally extensive spaces for the pliable forms of life that people use to make rules and norms and other means of extending the world."[51]

Berlant acknowledges the need to change structures and frames—but also that we live within these structures—and like the toxic infrastructures discussed above, they cannot be removed, easily repaired, or replaced by an idealist commons. They also tend to reassert themselves. Through engaging with infrastructure as the patterning of form and acknowledging our resistance to both these forms and changing them, Berlant allows for opportunities and change at the level of ideas and frames:

> **What remains for our pedagogy of unlearning is to build affective infrastructures that admit the work of desire as the work of an aspirational ambivalence. What remains is the potential we have to common infrastructures that absorb the blows of our aggressive need for the world to accommodate us *and* our resistance to adaptation and that, at the same time, hold out the prospect of a world worth attaching to that's something *other than an* old hope's bitter echo.**[52]

This is ultimately a constructive and inflectional understanding of infrastructure as a way to imagine the world—somewhat akin to Sara Hendren's discussion of reimagining the "broken world" through seeing differently (interlude, this chapter)—which is attentive to infrastructure as a framework (the protocols, how we process objects), advocates meaningful shifts of infrastructure, and suggests that our lifeworld can be shifted.

Reframing Academic Events (A Case Study)

Conferences, workshops, and symposia make up an important part of knowledge production, particularly in academia.[53] Academic events can be regarded as infrastructures for knowledge production. They are complex infrastructures but also relatively contained, which makes them a useful case study for considering how frames shift and can be shifted. The case study is, at heart, also about designing humane infrastructures. How can we make academic events better?

Academic events are characterized by conditioning and conventions, such as the program format, types of slots, management of Q&As, choice of sites and venues, use of space and presentation technology, and mechanisms for soliciting and selecting speakers and participants. There are also clear expectations of what an event should be like. For example, conference participants take their seats at a conference, remain seated, and expect mostly to listen, whereas keynote speakers typically expect to be introduced generously, stand up at the front of the room while talking, have a projection screen behind them, and be given more time than other speakers.

Academic events can be described as hierarchical, ritualistic, and structural.[54] These structures are not necessarily transparent, and as noted, there is a strong template (space, program, roles, format, turn-taking, etc.). Challenging such structures is often difficult. Feminist scholar Sara Ahmed outlines an exchange in which someone poses a challenge about an event only having white speakers:

> **You say: this event has a structure. The response becomes: this is an event, not a structure. It is as if by describing the event as having a structure, you are imposing a structure on the event. Even by describing a gathering as "white men," you are then assumed to be imposing certain categories onto bodies, reducing or failing to grasp the heterogeneity of an event; solidifying through your own description something that is fluid.[55]**

Ahmed describes an "institutional brick wall" and structures that many have an investment in not recognizing.[56] Through the above infrastructural inversion, she brings such structures in focus (chapter 3). Following Berlant, it can be argued that "infrastructure" provides a useful place for doing such uncovering and logic-shifting work.[57] An infrastructural reading of events would reasonably also include other categories and perspectives—some of which would be material—that can help decipher events as structural.

Another structural property is the substantial environmental footprint of many academic events, as they are mostly predicated on people traveling to sites.[58] Changing such patterns once traveling has picked up in a post-pandemic world is a slow process. Furthermore, whether there is a need to travel or not, there is typically a cost for attending conferences, which is easier to handle for well-established participants than for early-career scholars. Even if an institution covers some of the costs, there are also secondary infrastructures in play, such as reimbursement systems. Not everyone can wait for months to be reimbursed. Even if such infrastructure seems set, it can be changed—for instance, through making fast reimbursement an institutional priority or through creating routines for the institution to absorb the cost of as many expenses as possible (flights, hotels, fees, etc.) directly.

The typical annual conference at a hotel or convention center is heavily templated, removed from any local university, and is part of a financial-logistical complex. A common critique is that the time allotted for Q&As and actual conversation is minimal.[59] Value is assigned to speaking time—having or grabbing more time is better—as well as to smart, ritualistic, or even abrasive comments rather than generous and constructive contributions. This tendency for one-way communication is reinforced by the use of standard venues with a designated speaker position and screen upfront. It is not that annual conferences cannot be wonderful occasions but rather that there are structural factors at play and possibilities for reimagining the structure.

I will now introduce two accounts of changing frames and building capacity in this domain. One is a personal practice-based account, and the other one describes the development of an annual conference series during and after the pandemic. The first account illustrates how alternative frames may not be there from the beginning but emerge over time through many small changes and experiments. The second account demonstrates how external circumstances can challenge core values—such as the primacy of physical presence for conferences—but also how frames are resistant and embedded in various financial and ideational commitments.

Account 1: Curatorial Practice as a Nomadic Frame

Pushing the protocol

When I started to organize events in the early 2000s, I knew that they gave opportunities beyond scholarly exchange, such as articulating an institution and bringing exciting people together. Most of the events we organized at that time were fairly conventional. The idea of distorting the format of a seminar series or a conference was not on the table. Nevertheless, there were some structural properties that mattered. The venue itself was one such property. We used an open lab space that did not look like a lecture hall. It did have seminar tables and chairs (hence "reopening" or relating to a convention), but it also had multiple large screens, a technology-rich performance space, couches, an aquarium, and colors that did not seem institutional.[60] There were often many simultaneous activities and a flow of people, which meant that the conception of the closed lecture hall was challenged.

We also made a series of small-scale choices over time. For instance, I invited a student I had met at MIT as a keynote speaker for a conference on language education. Years later, we convinced an external funding agency to shorten the time for keynotes (they insisted on full-on keynotes) to allow for early-career scholars to get more time. I also thought it was important not to make the early-career scholars comment on the keynote speaker. Rather, they were encouraged to present their own perspective in relation to the theme of the session. One of the keynote speakers tweeted that "I take it as a v. good sign that over the last 2 days I find myself wishing that the 'jr.' scholars were doing the keynotes."[61] Making these events work was a deeply collaborative effort, and I relied on a highly skilled team in HUMlab. They were as engaged with the minute details and experimental setups as they were with welcoming our visitors and showing genuine interest in their work.

Another early intervention was that I would sometimes forego senior scholars who always spoke and took their place for granted at the discussion session at seminars and rather approach students or early-career scholars. This came from noticing that the people quickest to raise their hands were often the more senior scholars, frequently seated upfront, and I felt the need to counterbalance this pattern without unnecessarily affronting the senior scholars. Sometimes, I would approach early-career scholars and students before the seminar and tell them I might come up to them during discussion to see if they had any comments or questions.

Both these examples pushed the protocol of conferences gently: the position and time given to the keynote and the logic of the "raised hand" list, respectively. Gradually, we started to introduce other "innovations." For example, I more or less stopped introducing speakers at any length, as I noticed that the introductions tended to reiterate hierarchies where the keynote speakers got the longest introduction by far. The information was generally available online anyhow as well as in the conference information, and I also felt it took time away from other things. Much later, I learned that one risk with this strategy is to "remove" the credentialing or title of early-career participants for whom it mattered a great deal. Sometimes, including the "PhD" or a brief biography can be very meaningful.

We experimented with different ways of bringing remote people into the space, resisting the prevalent tendency (at least at that time) to put remote participants far down in the program. At one event, we started with wheeling in a remote participant on a mobile screen right after the introduction, doing the opposite of hiding away the speaker. Sometimes, we had remote participants

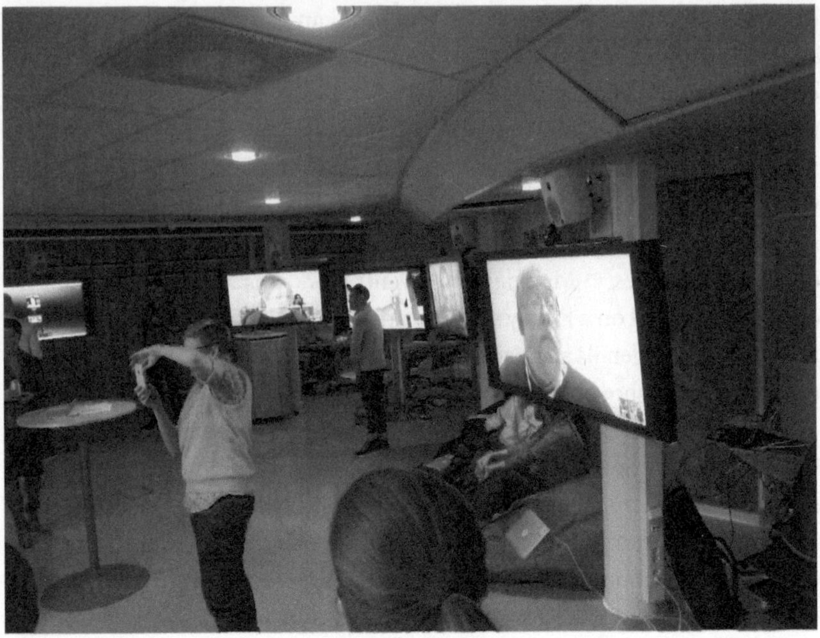

Figure 4.2
From the "Sorting the Digital Humanities Out" workshop, Umeå University, 2013.

on several screens in the same space using localized audio (figure 4.2), or even—as part of an installation—someone on twelve screens at the same time (one live and eleven recorded versions).

In general, we thought about how to start events and setting the tone differently. This is why we introduced initial conversations in the program—often extensive—where my leading question was "Why are you here?" I would have some things in mind for the session and would have talked to some people beforehand, which meant that the session would be semi-structured and curated but also dynamic and on the spur. Ideally, such conversations would create a sense of a comfortable, supportive, and dialogic space.

Manifesting values

When I relocated to the US in 2015, I was hesitant at first about organizing new events, as I associated my practice with the people, space, and infrastructure (capacity) at the Umeå lab. When I started again, the infrastructural context was different, but many of the principles and values that we had developed over the years turned out still to be relevant and usable. Also,

Figure 4.3
Workshop "Making Change through the Humanities: Institutes, Ideas and Infrastructures," KTH Royal Institute of Technology, 2018. Marisa Parham speaking. Photo credit: Francesca Albrezzi.

because I had access to less and different infrastructure, I had to do things differently. Together with colleagues, I carried out experiments with the format and became increasingly interested in securing even more time for conversation.

For example, I introduced "pair conversations" at an event on critical visualization in New York City (2016) where two people have a dialogue around a topic where they do a brief introduction but most of the time is open conversation. We stacked several pair conversations in the same time slot and made it clear that it was perfectly okay if some conversations were brief and others longer, resisting the notion that more time is better.

At UCLA, I initiated a seminar series with four or five people in an open conversation without slides. The first one was on encoding diversity and was rich and engaging if somewhat challenging for the conversationalists. Gradually, the action moved away from upfront to the rest of the space, with me moving between people and groups, which meant that the ideal space turned out to be a large flat space with round tables (see figure 4.3 for an example).

The framing of academic events is about not only the program, space, and presentation technology but also many other factors. Some of these are naturalized discursive patterns of the kind communication researcher Karen Tracy analyzes in her scholarship on colloquiums and academic discourse:

> **By and large, I would argue, faculty members fail to recognize the full array of ways in which higher institutional rank privileges them and creates disadvantages for graduate students. Because most faculty members are confident that they do not, or would not, abuse power, they presuppose that in a situation in which they speak, the merit of what is being discussed is the only issue.[62]**

The sense that faculty will not abuse power can be extended to faculty whose work considers "power" critically but still may not enact that knowledge in their own practice. Ahmed invokes "critical sexism" and "critical racism" to describe sexism and racism reproduced by people who believe that they are too critical to reproduce sexism and racism.[63]

The seminar chair, moderator, or convener has to lead by example, be clear about the values and rules, and be willing to act as needed. This is an individual as much as institutional responsibility, which needs to be agreed on and taught and where there should be support and mentorship. When I look through my scripts for various events, there is almost always a few things to say about basic commitments and conditions when opening the event, including "commitment to many voices, generosity, niceness and sharpness," "sharing time responsibly," "I will moderate accordingly," "focus on conversation, no slideshows, use of space" and "things and people are in place I think, and the rest is up to all of us today and tomorrow. Great if we can help each other out!"[64]

Such instructions are, in fact, about establishing or shifting frames. For example, the notion that conference attendees are "consumers" of an event can be resisted through telling people at the beginning that they are an important part of the making of the event or even that help in moving furniture or recording video would be appreciated. Stressing generosity as well as sharpness makes clear that the event in question will likely not be a "mean" one.[65] The way these things are presented matters—for example, being inviting and encouraging often helps, just like a degree of friendly or even humorous firmness when required, especially in relation to the everyday process of participating in a conference, but of course, any serious oversteps should be called out strongly and directly.

Expressing values and commitments is important but will not matter much unless they are manifested through the program, people invited, and moderation. If multivocality is important, organizers must look beyond the typical invitees, which requires a real effort beyond "token" or comfortable alternative representation. If it is important to include early-career scholars and experts, this must be manifested not only through whom is invited and supported financially but also in how the programming and "content" are shaped. If resisting or not reinforcing hierarchies is a priority, this should be clear from the organization and tonality of the event as well as in every step of carrying out the event.

The overall argument I want to make is that while I never thought that we would upend the format or change things radically, there was a gradual buildup of small things that helped me (and others) to rethink what an academic event could be, ever so humbly. In retrospect, it is easy to see how these small innovations had anchorage both horizontally and vertically, but it would have been difficult to articulate a direction at an early phase. It took many years of collaborative practice, experiments, and critical reflection to shape what might be called a style or a shifted frame—in fact, capacity to convene.

The resultant frame, however, should not be thought of as stable but rather depends on new innovations, experiments, and traversal configurations in what might be called a nomadic practice.[66] The commitment to values and incentives such as conversation, experimentation, and fairness is not restricted to specific formats but rather lives in the continuous challenging and realignment of the frame.

Account 2: Pandemic Conferencing

The annual conference of the Society of Cinema and Media Studies (SCMS) has been running since 1960.[67] SCMS is a mid-sized scholarly organization, which is also reflected in the size of the conference. It is large but not enormous like the Modern Language Association conference. The SCMS webpage on the conference features a photo of a conventional conference setup with a speaker and slideshow upfront in what seems to be a hotel or convention setting.[68] The physical setting as a key commitment will be the focus of this account, as this framing element came under considerable pressure through a global pandemic—COVID-19—that made traveling and on-site events practically impossible.

The SCMS conference is one of many annual conferences organized by scholarly societies as a keystone for their academic and financial activities. It is a particularly good example because it is exemplarily documented and because the organization has been remarkably open and transparent in their negotiation of the pandemic situation. While there had been suggestions to introduce virtual participation and digital platforms for the SCMS conference before 2020, there had been little accommodation.[69]

The policy on virtual participation is crystal clear in the 2019 conference program:

> **The SCMS annual conference requires physical attendance by all presenters.**[70]

Two main reasons are given. The first reason can be described as ideological, as the conference "places a premium on face-to-face interaction and conference-wide participation."[71] The other main reason has to do with the hotel venues and their infrastructure (or rather lack thereof): hotel technology is costly and unreliable. This argument essentially equates the organization and conference infrastructure with hotel infrastructure, and this infrastructure clearly impedes the organization's ability to facilitate virtual participation.

When the pandemic hit the world after 2019, SCMS and other scholarly associations were presented with major challenges at a time when other things mattered much more. The SCMS conference usually takes place in March or April, which gave the organization very little time to react. SCMS took the decision to cancel the 2020 conference, and in their motivation, they essentially declared that they could not do virtual conferences or help facilitate scholarly conversations in other platforms.[72] This is explicitly an infrastructural argument.

Again, it becomes clear that the conference infrastructure is displaced and outsourced. Whether hotel infrastructure or online platforms, this is not something that the association has full control over, owns, or imagines. It would have been surprising if SCMS, as a relatively small organization, had been able to come up with a virtual replacement at short notice, but it can nevertheless be maintained that they were also not prepared because the idea of virtual participation had not really been on the table—not part of the frame—and because their conference model is deeply entangled with hotel infrastructure.

Pandemic conferencing

The 2021 conference was made virtual through Zoom/Pathable with a program like those of the pre-pandemic years. The main difference was that the number of concurrent sessions increased, which speaks to a backlog but also shows that the virtual conference raised more than enough interest. It can also be argued that the fact that the format of the conference is conventional made it easy to transfer to a platform such as Zoom, which, in many ways, replicates a conventional event setup, although with an emphasis on the video (coming from a strong lineage of video meetings). Indeed, Zoom scales more efficiently than a hotel venue, which makes it easier to increase the number of sessions.

After a great deal of deliberation, the 2022 conference was divided into a virtual preconference and a physical main conference (four days instead of five, thus responding to previous critique of the conference being too long and expensive but also accommodating COVID-19 concerns).[73] However, given the surge of COVID-19 cases, the organization ended up having to go for a virtual conference. The decision was delivered in a video recording that demonstrates some of the tensions and considerations at play. It is clear that a driving force was financial concerns and multiple-year commitments to hotels.

There is a change in tonality, however, and there is now more openness to virtual platforms—probably as a result of the 2021 virtual conference experience and from pressure from the community. The organization talks about how the virtual harmonizes with the aspirations of the association to be an inclusive, international, and intergenerational organization.[74] This could have been a driving force for the conference, but at the same time, long-term alignment with a virtual model is clearly a very hesitant move. For example, the term "robust venues" is attributed to physical meeting venues, presumably the hotels (despite their "weak" infrastructure, as identified by the organization itself). The basic stance is that it is difficult to replicate the intellectual and social benefits of an in-person conference. Significantly, the video ends with bringing up the plans for the in-person 2023 Denver conference.

The 2023 SCMS conference was planned as a physical conference but also included a virtual preconference. In a members' presentation in 2022, some of the concerns were outlined.[75] One central factor was that the organization has hotel commitments until 2026, which means that the organization is locked down for four years and that the 2027 conference is the first opportunity to change the model substantially. It is easy to see how financial, infrastructural, and ideological arguments overlap. The hotel model, tightly connected to physical co-presence and renting out rooms as well as providing venues, restricts infrastructural and ideational maneuverability. Also, many members are used to and value the conventional model with on-site conferences in cities such as Chicago and Montreal.

The pandemic contributed to shifting the frame of the SCMS conference from not allowing virtual participation at all to becoming more open to virtual modalities. This may not be surprising, but what is interesting is how the resistance to virtual formats is embedded in infrastructural commitments and the financial model as well as driven by ideational considerations. There was certainly a shift and adoption of a virtual conference model, but seemingly more as an exception than a long-term transformation. Nevertheless, while the preconference virtual session in a post-pandemic setting is a limited move, it does represent an infrastructural and ideational modification of the frame.

Reconsidering formats

SCMS now has time to consider their annual conference beyond 2026, which was on the agenda for the association. One key challenge (unless staying with the current model whether on-site or virtual) is how to imagine what they want the conference to be beyond their current infrastructural and ideational entanglements. Moving forward will require negotiating the multiple forces at play to further modify or reimagine the framing of the annual conference. The association will have to come up with a viable conceptual, financial, and infrastructural grounding for the conference series.

While we cannot know the outcome of such a process, it may be useful to speculate about possibilities in order to consider how frames can be shifted practically in relation to this scenario. I will briefly consider two possible shifts for the SCMS conference: (1) opposing the one-site model and its reliance on not blending physical and virtual participation, and (2) reconsidering the structure of the program to support more dialogue and experimentation.

Moving from a single-site model can be done in several ways. One relatively conservative move would be to keep the single-site orientation but make sure that it integrates remote presence in a nondiscriminatory and creative way. There are plenty of researchers in the community interested in such issues who could help implement such a shift. This setup might not be achievable at traditional hotel venues, or at least not at *any* hotel or convention venue, which means that sites would have to be selected in a different way. Furthermore, SCMS would likely have to establish a stronger infrastructural setup themselves as well as knowledge about infrastructures (which can partly be found in the SCMS community).

Another more radical version of opposing the single-site conference model would be to give up the idea of a single site and adopt a decentralized approach. It would be possible to stage a multiple-site implementation where not everyone attending physically would have to go to the same site but rather could choose from, for example, three or four sites. Remote participation would have to be accommodated under this model just like in the more conservative version sketched above. This approach would reduce travel costs and the carbon footprint and possibly also open up a different sensibility where SCMS conferences would bring together a number of sites meaningfully (accommodating different communities, places, foci, etc.). The reduced size of the individual sites (even if there was a main one) could also make it possible to localize events differently—for example, at universities rather than at conference hotels.

The other shift considered—moving from a largely presentational format to a more dialogic, interactional, and experimental format—would require changing the program structure and the routines around facilitating events. The simplest option would be to increase the time for Q&A sessions—perhaps better called "discussion sessions"—and be very strict about timekeeping to safeguard time for discussion. However, this would reduce the number of slots in total, which might be a challenge for the association. Even if some time could be taken from elsewhere (keynotes, presentation time), the effects of such a change would probably be limited if everything else was kept constant.

Another option would be to make panel discussions, which constitute a substantial part of the SCMS program, much more dialogic. This could be done through moving away from the current model—several papers being read with little (and always vulnerable) time for discussion—to a model where there were initial brief statements from the panelists and ample time for dialogue. This model would not have to preclude individual papers, which offer an opportunity for an extended argument and serve a credentialing function, as these could be made available beforehand on the conference website or elsewhere. The statements would then relate to the papers and the overall theme but could have more punch, emphasize trajectories and intersections across the panelists, or point to some critical issue.

It would also be possible to change format types and include ones that focused more on exchange and dialogue, such as the pair conversations discussed above. As for other experimental formats and practices, it would seem that SCMS has the DNA and community to become much more experimental in relation to their own conferences, bringing together critical and practice-based commitments of the disciplines in question and also encouraging a broader development in an experimental-critical-constructive direction.

Because scholarly associations are good at community work and at developing mechanisms (building capacity), for example through working groups, mentor networks, and policy documents, they seem well positioned for thinking through, shifting, and sustaining new models of coming together to do scholarly work.

Arenas for Transversal Knowledge Production

This book has emphasized how knowledge production is conditioned and constrained by epistemic traditions, infrastructural templates, and organizational structures. As we have seen, systems of knowledge production are also enabling, both in their current form and in their potential, future forms. A recurring theme among initiatives and approaches surveyed has been critiquing and challenging the predominant model of knowledge production. This critique applies to the humanities as well as the university as a whole. One central argument is that we need to move beyond existing structures—such as conventional disciplines, individually based scholarship, and distanced modes of critique—to handle the challenges and issues we need to tackle as scholars and a society. The importance of interdisciplinary practice and world-facing engagement is often stressed.

There can be no doubt that we must revise the disciplinary structure and associated infrastructures to accommodate such needs both within and outside the academy. This situation, however, does not imply that the only option (if at all possible) is a complete overhaul or reset of the system—such as becoming completely post-disciplinary, disregarding earlier structures, giving up "irrelevant" topics, or equating the humanities with emerging humanities. Rather, I argue that we can be radical and frameshifting at scale without tearing down existing structures as a prerequisite, instead focusing on places in between and what can be contributed from different actors.

Such meeting places must support work across practices, traditions, and rationales in order to build capacity to mobilize humane and humanistic knowledge to respond collaboratively to challenges and problems. There must be a way for different stakeholders—including universities, corporations, civil society, and communities—to come together and work together beyond individual projects. In earlier work, I have discussed how such boundary work can be carried out across epistemic traditions and what qualities and processes can be central to facilitate dynamic meeting places, including integrity, curatorship, and genuine interest.[76] Through employing the rubric "arenas for transversal knowledge production" for such meeting places, I emphasize their heterogenous and interconnective qualities across knowledge traditions and infrastructures.

I use the term "arena" analytically because it has a generic and abstract quality that makes it seem less materially and institutionally specific compared to names often used for liminal formations—such as centers, labs, ateliers, studios, workshops, programs, meeting places, and institutes—all of which have their different valencies.[77] An arena in this sense is not an empty place or a blank slate—indeed, it does not even have to be a place—but needs to be meaningful and have some kind of direction. Direction is simultaneously a curatorial responsibility, collaborative effort, and an emergent property. Ideally, this directionality and sense of identity can be decoupled from the territorial ambitions typically associated with institutional efforts. This means challenging a model often predicated on privileging institutional achievements, taking credit, and owning the issue or activities.

"Transversal knowledge production" references work in such arenas that does not just comfortably extend existing traditions but that crosses boundaries in different ways and that has the potential to change positions, modes of engagement, and ways of thinking about the world. A transversal line is one that crosses two or more lines at different points. Knowledge traditions need to meet thickly and not in point.[78] Participants also need to meet across different modes of engagement and expression, including writing, reading, talking, experimenting, critiquing, performing, tinkering, prototyping, simulating, and building. There will have to be mutual respect, intellectual generosity, and committing to sharing on the level of both the institution—manifested through the arena—and the participants.[79]

I once talked to a colleague whose department at Umeå University was a result of a merger of many departments, each representing a sub-discipline within the same field. When the merger was done, a large café-style space was created as a place for new connections and ideas between the former departments. My colleague was skeptical about the café being sufficiently meaningful as an arena. Some people did not go there in the first place, people tended to sit with people they already knew, and not everyone wanted to mix their coffee or lunch with institutional engineering. Also, it was made difficult to bring outsiders into the space. Encouraging in-group collaboration—in this case, within the new department—can, in fact, result in discouraging connection making outside the core constituency. This is not to say that connections were not made or that there was no serendipity but rather that it helps to have something to gather around that feels meaningful and not contrived.

On the other hand, an arena that is overly instrumental or programmed might not be conducive to emergent ideas and a sense of creative freedom and space. Frameworks such as design thinking card decks, Post-it notes, strict business plan-logic, and step-by-step guides can seem contrived and heavily conditioned as well as set in a particular mode of thinking.[80] It is not that such tools and frameworks cannot be useful in context but rather that arenas need to be bigger as well as more open and multivalent. People should not be dragged into a rigid set of processes that are part of some compulsory model of change or innovation.

The level of instrumentality suggested or invoked is thus a relevant parameter when setting up arenas for transversal knowledge production. Being instrumental is important, but being too instrumental can result in a routinized approach that cannot accommodate different ways of approaching issues or allow for emergent and unexpected collaborations. Not being overly instrumental, as in the case of the café for the new department, can lower interest and incitements.

Both positions risk the collaboration and heterogeneity that Henry Collins, Robert Evans, and Michael E. Gorman associate with fractioned trading zones—the type of trading zones most likely to dynamically harbor boundary objects and interactional expertise.[81] A key difference between trading zones, pioneered by Peter Galison to explain how distinctive research communities in high-energy physics coordinate their actions, and the transversal arenas under discussion here is that the latter must be flexible across a wealth of communities and initiatives over time.[82] They also embed a very broad sense of knowledge production and stakeholders.

So, what will people and stakeholders gather around? They may be inclined to come together around a problem they are facing (e.g., recontextualizing climate change models in actionable ways) because they see an opportunity to reach further together and learn from others. They may gather around a research theme (supply chains or scale) relevant to all the actors in different ways, or a program where they get to work with students from fields outside their own—for example, exploring world views in city apps together with digital humanities students and the unhoused community. They may gather around tools and methodologies (critical visualization, infrastructural inversion, or terraforming), challenging large ideas ("the planetary," "AI," or "the carbon footprint"), a theoretical perspective (the posthuman), or an infrastructure (an environment for prototyping social forms or provoking algorithms) that does not exist elsewhere. They may gather even if there is no obvious alignment, but because there is enough trust in the arena so that stakeholders and participants come there to "be," engage, and experiment.

Most of these examples can be thought of in terms of boundary objects: things and concepts that attract participation across constituencies through being meaningful, malleable, and negotiable.[83] Arenas for transversal knowledge production are in the business of soliciting, suggesting, and negotiating boundary objects, but again, there should be an openness and ease to the operation. A low-threshold entry point can be to provide different types of programmatic activities and events, such as workshops, talks, and dialogues. These initiatives can be dynamic and unassuming, building on existing and planned work. This allows them to naturally ease into extended thematic engagements, collaborations, and actions on various issues, such as visualizing absent data, the conceptual and environmental footprint of generative AI, preparing society for future pandemics, climate narratives, or stress response modeling. Addressing these topics would require engaging across a range of constituencies, domains, and epistemic traditions.

Transversal knowledge production is about not only concrete results but also the process and the expertise it takes to support and facilitate liminal work. For example, research on interdisciplinarity points to the importance of catalysis, leadership, active management, and continuity in such processes.[84] This knowledge is often tacit, implicit, and reflexive, as we could see in the opening discussion in this chapter (a conversation on shifting sociotechnological systems). Arenas for transversal knowledge production can ideally help aggregate, discuss, and activate such knowledge over time—build capacity—across projects and initiatives. In terms of capacity, rather than building new shiny centers where people and resources are put together, there can be models where action and infrastructure—as well as trust—are distributed and situated to a significant degree.

Cutting across

Operating such arenas—at least on some scale—requires facilitation or curatorship and, as noted above, directionality. They need to be more like the type of humanities centers that have distinct leadership and positionality than centers that function as administrative centers or project hotels (chapter 2). They have to be driven by dialogue across communities and issues and must combine overall pluralistic directionality with a certain degree of integrity for specific projects, tracks, or cases.[85]

This negotiation can be facilitated and supported by curatorial practice. Curatorship is a notion based in the museum and art world, where it has expanded from an object and collection-centric role to a more creative meaning-making role.[86] It is nowadays used broadly in different contexts. I use the term "curation" to describe the job of supporting meaning-making activities between different interlocutors in an arena as well as in shaping capacity building.[87] Examples include the collaborative carving out of directions and boundary objects in an arena and the way that toolboxes (e.g., methods), infrastructure (space, platforms), and programming (activities, themes) are put together to build capacity. I argue that arenas for transversal knowledge production need curatorship to maintain their liminal position and direction across activities and projects. Such curatorship must have integrity and be built on mutual respect. As I have stated elsewhere, it should not be based on a high-priest model of curation or an overly humble and servile model but needs to be somewhere in between these polarities.[88]

Working across stakeholders and knowledge traditions is complex for several reasons. It is genuinely difficult to break free from the predispositions of what one does every day, our institutional context, and the way we were educated and professionalized. Breaking free also means not just extending existing networks but also genuinely searching elsewhere. It is worth repeating Anand's advice: form "uncomfortable relationships."[89] There is much path dependency that extends across different registers, levels, and infrastructures. For example, keycard systems may not allow you to casually visit another department, communities may have near-permanent distrust for other stakeholders, even middle management may not be able to allocate an hour or two for an "unbillable" meeting, an artist may not feel comfortable meeting with corporate CEOs, heads of organizations may not be prepared to forego the standard pitch and institutional template, there may be little willingness to take risk, development projects may impose unwanted abstract semi-technocratic models on industry partners (such as "clusters," "smart innovation," or "missions") or construe communities as passive recipients of knowledge, and administrative payment systems may not be able to handle payments or reimbursements to people without bank accounts or an address.

Transversal arenas have to relate to institutional and infrastructural logic as part of their daily practice—their own but also typically that of many other institutions, communities, and infrastructures. The complexity increases, since interests, incitements, resources, and expectations often vary considerably between partners. A simple but telling example is that people from universities and the public sector often presume that corporate actors, communities, artists, and other potential partners are willing to take time and make the effort—including sitting through long meetings—even though their reality is quite different from the academic world.

Nevertheless, universities have qualities that make them suitable for arena and capacity building, including that looser conception of time compared to some other actors. At the same time, universities are solid long-lasting institutions and so there for the long haul. Importantly, they typically operate across a very broad range of areas and interests, which means that they are versatile in relation to different needs and contexts. This breadth—especially at comprehensive universities—is complemented by profound and sometimes hyper-specialized knowledge in a range of fields. Furthermore, although universities are insular in some ways, there is often a willingness to engage as well as a certain degree of approachability and openness. Universities are not only there for themselves but also part of something larger. This connectedness and connectivity are perhaps most strongly manifested through the flow of incoming students—a constant reminder of that societal role and place.

Convening power

Universities are naturally not the only possible site for arena and capacity building of the sort discussed in this book, and they are not in any sense perfect. On the contrary, there are many problems with higher education—including obvious privilege and separation, insularity and aloofness, heavy administration, and inflexible organizations—but at the same time, there is also potential and hope, even if not always fully deserved. An important factor is that higher education has trust—not everywhere and always but more than many other institutions and actors.

The factors listed above—trust, comprehensiveness, openness, societal anchorage, approachability, and new generations of students—contribute to making universities into strong platforms for outward engagement, giving them what may be called "convening power."[90] Convening power can be ascribed to organizations such as the UN and the World Bank, and the term is also used for specific contexts, including ethics consultancy and food policy work.[91] I argue that universities have a crucial opportunity to convene across constituencies more decisively, not by themselves but together with others. Convening power is the ability to host and invite diverse participants (people, organizations, stakeholders, etc.) around relevant topics and for those invitees to be likely to accept the invitation and participate constructively.

Convening power is not just about single instances of convening but also about trust and opportunity built across time and modalities. Stockholm's convening power for environmental policy was strongly manifested through the 1972 United Nations Conference on the Human Environment, but as environmental historian Sverker Sörlin demonstrates, this outsized power was established and sustained through many factors, including science resources in Sweden, the work of individual researchers and politicians, the prestige of the Nobel Prize, and continued engagement with these issues after 1972 in various international contexts.[92] Convening power is thus a long game. Trust is built and maintained, interest is harvested, and there is a sense of moving forward collaboratively on important issues, not only discursively but in terms of action. Broadly conceived, transversal arenas and humane capacity need to maintain and build that trust and power across different challenges and constituencies.

The very notion of convening power indicates that there are power differences. I just posited that universities have more convening power than many other actors, which means that these nonuniversity actors presumably have less such power. Furthermore, some universities clearly have more convening power than others through prestige, location, networks, infrastructure, support, and resources. There will also be power differences among stakeholders and participants in any collaboration, project, or convened activity. Convening power comes with considerable responsibility with regard to power asymmetries. For example, differences in privilege should be minimized or even reversed, not increased; the convener should not be overly servile to "important" participants; and the convening institution should be careful not to promote themselves at the expense of other stakeholders.

In summary, arenas for transversal knowledge production must be pragmatically instrumental, comprehensive, epistemologically multivalent, and directionally meaningful. They draw on convening power and build on curatorship. Importantly, they are not meant to be frictionless, unconfrontational, or homogenizing but rather interested in the meeting, grappling, and clashing between cultures that Mary Louise Pratt associates with "contact zones."[93] Such work must be constructive and sensitive to power asymmetries and not get stuck in one perspective or theme.

The Infrastructure Clinic

The Infrastructure Clinic is an attempt at creating humane and humanistic capacity that has been shaped by a range of conversations and contexts over a decade—some of which are described in this book—but was principally imagined and sketched out by Sha Xin Wei and the current author in a focused multiple-year conversation. These processes brought together different experiences, disciplinary positions, and types of transversal knowledge production, not least from humanities, mathematics, art, and media technology. The IC provides *one* example of capacity building along the lines discussed in this chapter and book, drawing on the critical, convening, challenging, and creative (or imaginative) powers at the heart of such mobilization.

A crit session or studio critique is a method often used in design and art education to discuss, synchronize, and assess projects—predominantly advanced students' projects developed over some time. There are different variants, but a common model is that invited experts and commentators critique the project, which is often represented by a material artifact, and engage with the designer or artist in an in-depth dialogue around their work.[94] Student peers and faculty will typically also be present. The artifact is not merely presentational but central to the argument and to the dialogue. Such sessions are ideally collaborative and constructive as well as critical.

Critique sessions are ritualistic and performative. While there can be a kind of idealistic shimmer associated with crit sessions—as inherently constructive, useful, and developing—they can also be problematic. For example, they have been critiqued for being one-sided, hierarchical, and unconstructive, although this is, of course, not always the case.[95] Regardless, it is important to be cognizant of the conditioning of the tool.

The Infrastructure Clinic is a reasonably fleshed-out model that has not yet been implemented at the time of writing. This means that it is a good candidate for a crit session. Dissecting something in the process of being made is quite different from considering an early-phase idea or a finished product. In the following, I will stage a simulated critique session of the IC as a small experiment in form meant to support a critical conversation more open and dynamic than a regular academic narrative, hopefully highlighting some of the choices and tensions at play.

"Simulated" indicates that it is not a real crit session but rather a staged one where the current author, on the one hand, has all the parts, but where, on the other hand, the session also draws on multiple dialogues and events where critique has come up and been discussed. The main thread consists of my own reflections on the conditioning and framing associated with the Infrastructure Clinic in the context of this embedding.[96] The questions and answers are mine but are also part of a larger dialogue (as also indicated by notes). Again, the IC as a concept is a joint venture between Sha Xin Wei and me. The respondent in the crit session, however, is me, not anyone else, and hence my responsibility.

At a crit session, the project would be present in the space. This space could be physical, digital, or mixed, but for our purposes, a mixed space makes most sense. We would have to imagine a representation of the IC in physical studio space, perhaps through projecting a concept description on a wall, key assumptions on other walls, and a schematic representation on the floor. A couple of prototype cases would also fill the space, such as physical models of possible infrastructures or of the operation required for a specific infrastructure (expertise, community engagement, components, frames, etc.).

Central concerns identified, such as scale, instrumentality, intellectualism, and solutionism, could be manifested through these models. For example, the issue of scale could be brought forth through juxtaposing differently sized infrastructures, ranging from the planet to small and local urban sites and embedded algorithms. In a more developed version beyond the scope of this imaginary crit session, there could be interactive installations such as worldview-dependent representations of infrastructures or adaptable algorithms running on databases.

Here is the concept description of the IC projected on the wall:

> **The IC is imagined as a world facing service to which people and organizations can bring some social form in which they live, like a city, a way of life, or a social or technological infrastructure for treatment and care. The IC team will work with existing centers and organizations to convene stakeholders, interpreters, knowhow, and sponsors to make sense of the situation, and come up with ways to collaboratively make the infrastructures work better, that is, in a more humane, efficient, and life affirming way. The response will be informed by humanistic and humane knowledge and shaped by the patient.**[97]

Often the dialogue in a crit session will be initiated by the instructors or external experts. This is the case with the current crit session.

> *Scene: The external reviewer stands in the midst of the infrastructure models and visual projections. The respondent stands close to the reviewer, leaning on one of the models. A group of students and faculty are seated on chairs placed at one*

end of the space. Some of them are sitting on big pillows on the floor. At the other end of the space, four screens on stands show remote participants. Placed close to them—on their own set of chairs—is a review committee consisting of five people. The external reviewer raises their hands and starts off the session proper.

Q: I am happy to be here to review and comment on the Infrastructure Clinic project. Let's get started at once with the first question. Why "clinic"? Clinics are associated with displaced locations, designated roles, authoritative treatment, a sterile and controlled environment, and irrefutable encoded expertise.[98]

The respondent moves slightly to face the reviewer, seemingly a bit surprised with the brief and direct introduction. Normally, there would have been a longer introduction before the first question.

A: Thank you. Yes, this is right, even if there are other kinds of clinics, some of which are closer to what we are imagining, such as Natalie Jeremijenko's Environmental Health Clinic and community clinics. One reason for our choice is that we needed something that was conceptually workable and somewhat provocative. Clinics are essentially a positive framing after all—people go the clinic to get better—and for good and bad, the clinic model is positivistic in the sense of being about solutions. We felt that it was important to complement and challenge the critical deconstructive tradition at heart of much work in the humanities, and the clinic embodies some of that tension, hopefully in a constructive manner. We wanted to push an instrumental sensitivity without succumbing to a solutionist model, extending the repertoire across a critical and constructive spectrum. We also hope that the tension built into running a humanistic and humane clinic will be productive. In this sense, the clinic is not a set frame not to be critiqued—it is what it becomes through our interaction with the model.

The reviewer takes few steps and points to one of the IC models—the one that shows the institutional placement of the IC in a university setting outside the regular structures, physically semi-detached, as well as a range of potential supporters from within the university and outside agencies.

Q: But is the IC not just a result of adjusting to external pressure from government and funders with a nominal interest in engaging the humanities, ethics, justice, and the like, from universities framing themselves in relation to global challenges, ethical AI, and planet-level care, and the odd university seeing themselves as a medical school for the planet? To be honest, the IC strikes me as an opportunistic and predictable endeavor.

A: I realize that it could be understood as a response to such pressures, and in a way, it is. We believe that it is a good thing that there is an interest in bringing in humanistic and humane knowledge, although we are aware that it is often given a peripheral or ornamental role and that there is internal resistance. This is why we need to be involved in shaping the response and letting our work, curiosity, and engagement be a driving force. And this kind of competence needs to have a place at the table, whether the topic is planetary health, developing a city site, or ethical AI, not just to provide context or as a token gesture, but integrated, troubling, and constructive. The IC is intended to help and work with patients rather than trying to respond to pressures. The patients will not be predictable.

Some in the audience seem taken with the directness of the question, whereas others appear amused or even approving of the aggressive stance. The reviewer nods but continues to look at the model in question and also the adjacent ones—all about the IC as an institution—and mimics searching for something.

Q: Okay, but what does this clinic look like? I do get a sense of the clinic from what we see in this space, but at the same time, it is mostly words and a few crude infrastructure models rather than a strong sense of the clinic itself. There is a schematic chart here, yes, but to me, it looks like something produced by a university administrator for a funding proposal or something. I do not really see what the clinic would be, what it would be like, what you would like it to be.

A: Well, I guess we do not really know yet, but we know what the clinic is not if that can be helpful. It is not a university center or institute. It is not a building or a space. Earlier in the process, we considered building a lab-like space, but it became increasingly clear that the clinic itself needs to be a distributed infrastructure—not only for working with others within the university but also for engaging with client or patient infrastructures outside the university. Also, the IC is not a think tank. We are interested in changing things for the better through being involved in processes of collaborative making, not through some academic think tank looking out on the world and coming up with abstract solutions. We believe that the university is the right place for this type of work, not one university but the university broadly speaking, but only if the operation can be truly outward reaching, respectful, challenging, and synergetic.

Q: Negative definitions tend to have limited value I find, but I get it: you think of the IC as sitting outside the conventional formats. I would have appreciated something more concrete though, but fine. But what is new and innovative here? There are countless initiatives trying to make the world a better place. I realize that we probably need quite a few, but I wonder what it is that makes the IC different and worthwhile? What do you bring to the table? Name one key innovation please!

The respondent has backed away from the reviewer somewhat, perhaps still shaken by the quick pace or the lack of more inviting and friendly questions. There is some defiance in the stance though, and their voice now seems to carry more strongly.

A: Right, we do not think of the IC as a traditional university-based initiative. Sorry about another negative definition, but it is *not* a center or an institute. One innovation is that we think of the IC as a service rather than as any of those and other institutional labels so common in academia.[99] There are several reasons for this move. Service as a category is not often used for academic endeavors and is not well defined in that domain, which means that it has a certain degree of flexibility and malleability. Service functions are much more readily associated with the administration and university-central functions, such as HR, development offices, and IT. The reach of such service functions—across the university—is an important quality to draw on for the IC as a service. In this way, the IC can ideally exist somewhat outside the regular academic territory and does not need to be locked to one discipline or school. The service framing will hopefully make it clear that the IC is not an end in itself but rather will be lightweight and premised on making things happen with others. Genuine engagement is central. It is important, however, to stress that the IC as a service does not equal a servile function driven by administrative goals. There needs to be directionality and an intellectual drive.

Q: My experience is that talking about collaboration, outside stakeholders, and the like is much easier than making such work happen. You talked about token figures earlier, and as someone from outside the university world, I feel that I have been a token person in many university-based contexts. I have participated because I have seen promise, and although they obviously needed and wanted someone from the outside, they have not been really interested in transformative engagement. This is how it has felt to me. I am sad to say that this has been particularly common with humanities-based initiatives. I do not think it is ill will or anything but probably a combination of traditionalism and not being used to working with external partners and actually letting those partners change oneself. What makes the IC different? What is behind your talk about making things happen with others? And where is the action and making? And if you do not mind, I would like your response not to resort to the kind of abstract institutional arguments you have favored so far in this conversation.

Although the sharpness is still there, the reviewer now seems more personal and, despite the final jab, in more of a conversational than a purely confrontational mode. A few of the faculty and students in the space are leaning forward in their seats, presumably engaged in the discussion.

A: Yes, I guess there are token figures in different camps. There is an institutional argument here: that the institution does not incentivize outward-reaching engagement beyond a few set models and that the structure holds people and ideas in place. The IC is an attempt to break free from some of these structures while also holding on to them—or at least holding on to the knowledge and potential that live in those structures. But let me not focus on the institutional. Beyond that, I think what we need is genuine engagement, dialogue, mutual respect, trust building, a willingness to learn and change with others, seeing, considering, and understanding different perspectives and positions, and arenas where some of the talking, imagining, and building can happen. Such arenas, which do not have to be physically based at the university, can be helped by curatorship and the convening power of the university. I know that there is a lot of interest, curiosity, and well-meaning across positionalities, but we need to work on the conditioning for making things happen in between. A lot of this also has to do with experience and time. We imagine the IC to work as an intermediary, interlocutory, synergetic, curatorial partner that can help facilitate conversations and work that might not happen otherwise and that can be useful when approaching situations like the ones you described—for example, humanities programs approaching outside partners—or when taking on a client or patient infrastructure. And it is about not just conversations but also concrete building and making in collaboration with others: critically infused making. The IC could well be involved in making algorithmic systems, planning and building urban sites, designing innovative university institutions, and devising schemes for small-scale local solar energy.

Q: I certainly admire your ability to stack adjectives and they do lend more heart and DNA to your project. I find that valuable if a bit long-winded. At the same time, I cannot help wondering about the size and scope of the IC. It seems so infinitely large. We have not talked a lot about the infrastructure part of the IC so far, but I find that infrastructure is almost anything these days. Is that not so? And the IC is not limited to any specific kind of challenges or problems? Is there any limit to the scope of the IC, and how are you planning to scale things? Where is the cut?

There is some laughter in the room. The answers from the respondent have indeed been long and wordy. Also, the reviewer's snappiness seems appreciated, especially when combined with a bit of warmth and what now seems like real engagement.

A: Yes, it is a big idea and enterprise. And infrastructure has become something to come together around. I guess one way of looking at it is that we are not trying to stake out a territory as much as traversing it. The cut is the intersection with other things—including people, centers, organizations, corporations, and city administrations—and we always depend on them. We do not—indeed should not—have to fill the territory ourselves. Without the other things, there is simply not much there. Well, there is an idea and a setup. The bottom line is that we believe that the humanistic and humane have bearing, power, and expertise across domains, and we are convinced there is a need for capacity that is generic rather than specific. This generic capacity is also needed to be attractive and comprehensible to the outside, whether institutional partners, local communities, funding agencies, NGOs, corporations, innovation systems, or infrastructure owners. Also, we do not know the needs of the patients and must be open to these through interaction with stakeholders. Such interaction gives specificity, just like perspectives, practices, and expertise from the network. Scalability comes through cases. The IC itself should be small and lightweight. Also, it needs to engage with collaborator, partner, and stakeholder infrastructures as much as possible rather than build new university-based infrastructures.

Q: The humanities are given a central role in the IC as a model. Do they know this? I get a sense that the IC constructs an ideal form of humanities—ideal for your purposes—that does not exist, perhaps does not even want to exist. Why would they be interested?

A: This is a good question, and there is certainly a risk of idealizing the humanities or reinterpreting them for our purposes. The IC may be reaching out to all humanists and all of the humanities as well as related fields, but it is not counting on everyone being interested. We believe we have a good chance of engaging with some of them because there is an interest in making the world better and we already have a network in place. We believe that people are looking for initiatives like ours. And we will be looking for them as part of our connective work, not just humanists mind you, but that is also the point. We realize that this will not appeal to everyone or to all of the institutional humanities, perhaps even as a matter of principle, but we believe things can change, and again, we do not need everyone. Also, it would be wrong to distance ourselves from the established humanities—some of the most important scholarly perspectives and tools come from there. One thing that makes us interesting, I think, is that our model is driven by the seriousness and edginess of critical perspectives, and we seek to instrumentalize those respectfully. We also have an idea of how to do this through methods, perspectives, theories, and values negotiated from scholarly work and other knowledge domains, which is a perhaps an innovation I should have mentioned in response to your earlier question.

Q: And conversely, what makes you believe that external entities and stakeholders would be interested in the IC? Why would they come to you? Why would I come to you? Also, what would make the university reach out? It seems like a closed-off world for which it is difficult to engage with the outside world profoundly, especially for fields such as humanities not used to having such contacts.

A: This is a real challenge, of course, but we have seen the IC as an outward-facing service from the very beginning. This is where the opportunity lies and also what is fun and important about the IC. This sentiment has directed the framing of the initiative. A clinic needs patients and cases, and a clinic is also decipherable to the outside world. Infrastructure is a kind of currency for the modern world and something to come together around. But much is about how the IC engages with the world—it must come from the heart, from difficult but exciting ideas and cases, and from a real will to talk across domains. And the cases themselves will make a big difference too, as these will provide a cut and a way to be specific and relevant. The IC operation also includes something that most clinics will not have—namely, an extensive event program meant to engage stakeholders; discuss themes, challenges, and research tracks; and come up with action plans. Some of these events will be high profile, whereas others will be small and targeted.

The exchange has been going on for quite a while now. People still seem reasonably engaged, perhaps helped by the relative sharpness of the dialogue, although there are some expressions of fidgetiness and tiredness. One of the senior faculty keeps looking at his watch. Some of the participants are looking at their notes to be prepared for the program following the initial review-and-discuss session.

Q: This is helpful, but how all this is going to work is still unclear to me. One thing that could help us better understand the potential and operation of the IC, however, would be to hear you talk about how the clinic will actually work and give us a few examples of patients or perhaps rather patient infrastructures.

A: The IC does not exist yet, and cases will be created with the patients. So, the following will be provisional. However, if you look at the chart—the one you said looked administrative—it shows three types of inflow of patients: general influx, outreach, and by referral. Once we get started, general influx is the stream of patients that we hope will come because they know about the Infrastructure Clinic, find information about it, or get a recommendation from somewhere. This will take some time to establish. Outreach is us and the network reaching out to potential patients or patients we have already involved in discussions. This also includes contacts made and patients discovered through events and other programmatic activities. Referral is a broad category referencing things—themes, questions, challenges—that arise in the IC network, involve stakeholders, and can be elevated to clinics or cases, as well as cases that come out of structural collaborations with partners such as funding agencies, donors, NGOs, and corporations. Once a patient infrastructure has come to the IC, it will be surveyed by the IC team together with key stakeholders—not only organizations but also people on the ground—and a small pre-clinic group will be formed to investigate the infrastructure and look at relevant research inside and outside the IC network, perspectives, questions, attached themes, and tools. After that, a case or clinic is started. As for actual patient infrastructures, we imagine these to be very different in terms of size and nature. As should be obvious from the rather large model over there, we think that the planet can be the patient. The lines of code you see next to the planet come from another example: they are a mock-up of scheduling algorithms for medical appointments. These have often been shown to be biased against people who are already underprivileged. Stakeholders for such a clinic could include patients, government agencies, clinics, and service providers.

Q: We are now coming to a close, and I would like to circle back to the first question. I asked about the authoritarian and expert-driven model of the clinic as a model and did not really get a response to that. After listening to you today, there are some things I would like to add to that question. How will you deal with differences in power and opportunities? How will you manage the balance between being based at a university and claiming to have a strong inclination to change things and perhaps even do activist work? And finally, how will you avoid the IC becoming seen as yet another center run by people with a pitch, grand and at least partly self-serving ambitions, and a top-down model?

A: Sorry for the incomplete answer. Yes, the clinic as a frame can certainly seem to invoke a hierarchical expert model: doctors at the top of chain and patients with little or no say. The IC is a different clinic in many ways, including catering to infrastructures as patients. We will push and negotiate the frame in different ways. In terms of hierarchies, we hope that the IC will be a collaborative, multi-perspectival, respectful initiative where everyone contributes to defining, handling, and owning the case in processes of critiquing, imagining, negotiating, and making. Participants, including patients, researchers, and stakeholders, are experts on different things. We need to manage situations with asymmetrical power situations fairly and constructively with an eye to participants with less power.[100] All this means that the clinic will have to be aware of its own conditioning and, as much as possible, make that conditioning visible and—when appropriate—recruited to help change and reimagine infrastructures. The IC is about changing things and thinking differently both about the infrastructures themselves and also about the world and its systemics. Sometimes, that work will have an activist slant, but it will have to be done through generative processes with other stakeholders, including ones that may seem adversarial. And relating to your final point about the top-down self-serving center, the model I have presented today is meant to counteract those forces. The curatorship, facilitation, and interaction with others make up the texture of the IC and will—must—be generous and collaborative. We will also do everything we can to live the values we talk about on a day-to-day basis: this is central to the whole operation. The IC must be a humane infrastructure.

The reviewer looks around, smiles for what seems like the first time in the review and begins to clap their hands, followed by the others, including two of the remote participants, one using the applaud button in the tool and another one clapping their hands in front of the camera. The respondent looks relieved more than anything else, but there is also something new in their bearing. They seem stronger. As we leave them, there is a short break before the open discussion.

What Goes into a Capacity?

Although the Infrastructure Clinic is only one possible instantiation, it illustrates the conditioning, framing, and convening that are part of humanistic and humane capacity building as modeled in this chapter and book. An underlying idea, directionality, and set of core values form a conceptual foundation that shape the frame, which needs to be both sharp and generic at the same time. The sharpness is necessary for generating interest and for making a clear case for the operation. Genericness opens up the operation for many cases and constituencies, required for a capacity for mobilizing humane and humanistic knowledge to respond collaboratively to challenges and problems. The frame comes with conditionality that must be considered, observed, and negotiated.

The concept and frame get manifested by an implementation, which, in our case, is not likely to be a conventional center, institute, or building but rather a transversal networked arena that is open and agnostic enough to accommodate different people and institutions, a range of scales and epistemologies, and a variety of cases, issues, and themes, while also providing concentration, "thickness," and productive and sharp "cuts." It also helps if it is as non-territorial as possible.

Curatorship and convening play a key part in facilitating the operation, helped by dialogue, curiosity, and a toolbox of instrumentalized critical perspectives, methods, and approaches. The tools, which serve as boundary objects, should preferably not be too specific or too large but rather applicable and negotiable. Concrete cases and themes shape the toolbox, but they are also informed by the tools. Large-scale and worldview-altering theories and frameworks can help provide a conceptual backdrop as negotiated. Knowledge is predominantly networked, and infrastructures are shared and situated across knowledge communities and constituencies.

Capacity building includes these and other components and structures. As noted, "capacity" seems like a more dynamic, responsive, and generative framing than "infrastructure." It is also (even) more generic, which can make it house a wider range of "things" and qualities. Another perspective on capacity building can be given through considering what things and qualities goes into such capacity.

What things and qualities? There is no complete or ordained list, of course, and any instantiation (such as the Infrastructure Clinic) will be different, but in the following, some of the key factors brought up in the book are reiterated:

1. *In between*: Create respectful, multivalent, and multi-scalar "spaces" or "contact zones" that are open as well as programmed and not territorial (build trust, encourage connectivity, support unexpected dialogue, be meaningful).
2. *How*: Be oppositional, edgy, and radical as well as generous, curious, and constructive; care about the value and potential of different types of knowledge, practice, and backgrounds; be reflexive, also of your own conditioning; learn from others.
3. *Networked capacity*: Work with partners in the humanities, design, science, art, industry, NGOs, government, and communities (and elsewhere; build genuine networks, not with everyone at once but gradually over time, shaping nimble, networked capacity).
4. *Critical tools as capacity*: Build a toolbox stemming mainly from critical perspectives and methods, negotiated and packaged but with regained edginess and power (not watered down, with integrity and bite, collective diverse capacity).
5. *Convening power*: Find ways of getting people, institutions, and other actors across domains to come together actively around themes and issues, always with an eye on the before and after as well as the present (building relationships and understanding, establishing convening capacity, preparing action, making things happen, having direction, supporting agency).
6. *Situating*: Reach outward (also outside yourself), be elsewhere, use distributed networks and infrastructure, be aware of asymmetrical power relations, build on mutual interest and shared potential (and consider different levels of engagement and criteria for engaging), trace and develop research themes and emergent "cuts".

7. *Operational capacity*: Establish operational practices and tools such as (a) experimental and generative programming (pulling together interests nimbly as well as long term, staking out new terrain), (b) double affiliation (to build bridges through faculty, fellowships, etc.), (c) distributed/situated/shared infrastructure (rather than necessarily building new infrastructures), (d) curricular development as a means to channel interest and make a long-term difference (caring about the students), (e) ways of supporting "liminal" careers, and (f) curatorial practice to identify intersections, cuts, themes, people, and challenges that can power the operation.
8. *Range*: Support diverse knowledge traditions and practices, by themselves but also as part of a place where they meet, clash, and necessarily learn from one another (thickly, across modalities and modes of engagement, critical and constructive at the same time, unquestionably intellectual).
9. *Direction*: Push and challenge conditions, frames, and infrastructures in academia and society as appropriate (in a humane, efficient, and life-affirming direction).
10. *Commitment*: Be curious, caring, generous, sharp, imaginative, experimental, and unafraid; manifest the ideational grounding of the operation.

Using the framework established in this book, capacity is the action-oriented set of resources, tools, networks, knowledge, and arenas required to form a packaging or frame that is decipherable to different actors, meaningfully manifests the values associated with the endeavor, and engages with different actors productively across a wide range of challenges and problems. Such capacity, when actualized, can ideally be a humane infrastructure used to critique, imagine, and design humane infrastructures together with others.

Conclusion: Toward Humane Infrastructures

This book details a journey driven by a commitment to effecting constructive change. A central question has been how to generate productive change collaboratively through critical and humane knowledge, especially when epistemological, institutional, and material obstacles stand in the way of meaningful ideas, values, and directions. How do we create humane infrastructures? The starting point was predominantly epistemological, relating to knowledge production and academic structures. One central concern was that research tools and instruments were often taken for granted and seen as neutral and predetermined, although they clearly have thinking or logic built into them, whether a search interface, database structures, a telescope, or a visualization wall.

Such "infrastructure thinking" is not only epistemological but also institutional and cultural. It turned out that research infrastructure as an emerging framework did little to alleviate this concern but rather reinforced the naturalized, instrumental, and objectivistic nature of infrastructure. It was also steeped in a science and technology model, even when applied to the humanities and social sciences. There was a worrying lack of critical reflection in research infrastructure work, also regarding humanistic infrastructure, and a strong sense of detachment, which seemed out of touch with the world.

Despite this detachment, the world kept creeping into the narrative, not just through the political and societal aspects of research infrastructure but also because it is generally difficult to separate academic and civic infrastructure. The infrastructures of the university—such as reimbursement systems, housing (or not) for students, politically supported investments in research instruments, and innovation centers—reach out into the world and vice versa. Housing and food insecurity among students on a University of California campus cannot be relegated to society or the academy but is necessarily part of both, which is also true of associated infrastructures. Another current example is when a government sees the need to securitize universities (see preface) or when politicians argue or legislate in favor of removing unwanted disciplines or specific course literature.[1]

The world is also affected by research infrastructure in other direct ways—for example, through the electricity consumed by storing visual data, maintaining large online archives, or carrying out terrestrial-level calculations. In an intriguing experiment, media artist and researcher Stephen

Cornford interrogates the environmental cost of using satellite imaging to survey the planetary surface in terms of how many pixels (the smallest unit in a digital display) can be created from each square meter of terrestrial resources, especially minerals.[2] In doing so, he inverts the notion of resolution, normally regarded as the spatial quality of the image measured in pixels according to a more-is-better logic (e.g., an 8K screen or a 60MP camera). He accentuates the physicality and cost of the pixel as opposed to seeing it as a numerical value or a bit of code. According to Cornford, images and pixels come from ground, and their production carries with it "environmental damages that should also be accounted for in the technical specifications of our instruments."[3] His thought experiment is an example of infrastructural inversion (chapter 3) used not just to change our "worldview" but also to suggest a different metric for devising and building tools or engaging in the imagery-rich infrastructure of the world.

It gradually became apparent how the story of infrastructure is embedded in a range of societal and cultural developments, including a rationalist strive toward standards, usefulness, and control, and how there were multiple overlapping lineages. For example, critical infrastructure is primarily driven by a vulnerability and protection logic with a long history, but it is also related to the idea of failing infrastructure that emerged in the 1980s. Research infrastructure, on the other hand, was driven not only by research—the need to fund and consolidate costly instruments for science—but also by multiple other factors, such as higher education as a global competitive market, universities as contributors to innovation, and national (and in some cases pan-national) pride.

Infrastructure does not just embed values and assumptions, it also manifests them. Think of a public bench designed to hinder laying down or sleeping, how medical scheduling software can be systematically biased against underprivileged communities, or what the detachment of research infrastructure tells us about its inflection and blind spots.[4] We rarely see that qualities such as inequity, care, and reflection are highlighted when things such as "smart cities," "3D printing," "generative AI," "augmented reality," or "cryptocurrencies" are presented. Infrastructure in the making—new or projected—is often shiny and full of promise but, at the same time, limited by the frame of infrastructure (big, costly, instrumentalist, technological, standardized, long term, benevolent, detached, etc.).

Through looking at how infrastructures have been constructed and negotiated historically, it became clear that what in many ways seemed stable, almost fixed, was not. The NATO Research Infrastructure Committee has benefited from the flexible and negotiable quality of the term (chapter 1). Research infrastructure as a composite idea and framework—now a far-reaching reality for many institutions of higher education—is only about twenty-five years old (chapter 2). The dynamic nature of infrastructure can also be evidenced in the dramatic expansion of its territory over time ("an infrastructural world"), supported by the elasticity of the notion as well as by what appears to be a lasting timeliness and usefulness of the idea.

Turning the Tables

If infrastructure was shaped and negotiated over time, can we not by the same token reimagine and reshape infrastructure moving forward? The first step for scholarly infrastructure had been to challenge assumptions such as the objective accuracy of visualizations (through visualizing uncertainty and asking questions about missing data), access as the end goal for digitization (through letting active use and specific research questions drive the interface or tool), and singular points of view (through facetted browsing and multiplex screenscapes). One emergent question was whether the worldly entanglement of academic infrastructure could also be made more active and directional. Furthermore, can we not shift or tweak the conditions that are often taken for granted (e.g., the big size of infrastructure or the central place of technology) or use inflection (such as "disability," "power," or "trouble") as an active force when designing infrastructure? Such a model implies that infrastructure work is about challenging and shifting situated (real) infrastructures and the frameworks associated with infrastructural systems as well as infrastructure as an idea.

Along these lines, Marie Louise Juul's resistance to the solutionist and disembodied ethos of femtech—in fact, proposing a set of alternative foundational ideas—operates both on the level of sociotechnological systems and on the level of actual design work (chapter 4, interlude). Matt Ratto's invocation of more-than-human perspectives for the addiction treatment program in the same conversation was about concrete design shifts (e.g., affecting the "speech" of the conversational agent) as well as a conceptual reorientation for technology work (challenging the "realist" notion of computer simulations). There is much to learn from these examples and more generally from work going on in liminal areas, including critical design, environmental humanities, digital humanities, disability studies, and food studies.

Similarly, if we believe that research infrastructure should have a stronger anchorage in the humanities beyond adapting the humanities fully to the existing frame (chapter 2), we will have to operate across different levels. We need to not only consider intellectually and methodologically worthwhile apparatuses (as discussed above) but also affect the idea and the framework through a process of negotiation and policy work, which cannot be done convincingly without an ideational basis. But at the same time, this basis cannot completely break away from the template of research infrastructure. For example, exploring the potential of the humanities beyond databases and tools in a Swedish working group lead to a formulation of a societally engaged type of humanistic infrastructure, although still mostly within the logic of research infrastructure.[5]

One of the steps taken in this collaborative effort was to suggest new categories for the Swedish Research Council definition of research infrastructure, including "experimental workshops" and "complex digital research systems."[6] The amended definition was adopted by the Research Council and other government organizations, at least in some documents for some time, including a report about the future research landscape sent to government from six state agencies.[7] Changing frames and framework is a long game, sometimes seemingly futile, but efforts to move positions can nevertheless be important.

Work in the world shows that shifting frames is possible, far beyond the importance of "research infrastructure." In Minneapolis, after the murder of George Floyd, the urban landscape—the infrastructure—changed irrevocably through what took place there and through extensive and difficult community work (chapter 4). The community also worked alongside the city to change the frameworks that condition infrastructure. As a result—however insufficient—the City of Minneapolis and their Department of Public Works have changed regulations, practices, and—at least on a discursive level and to some degree—the vision of the city.

All the things described above meshed with an increasing—eventually unavoidable—engagement with the societal challenges and complex problems we are facing as a planet and the crucial question of how critical and constructive work can help us generate responsive and responsible power. Such work cannot be done only on a case-by-case basis or through single themes, however important. It needs to have a systemic and generic character as well as recognizability and reach. A central question became how critical knowledge can be instrumentalized responsibly and meaningfully across domains, resulting in toolboxes that can be part of a more general capacity building (chapter 3).

The next step was then to consider what goes into imagining and implementing such capacity, including the necessity of roles and operations that work across actors, stakeholders, disciplines, types of knowledge, modes of engagement, and themes. The collaborative investigation of such liminal perspectives from the point of view of institutions, fields, and people led to the articulation of arenas for transversal knowledge production (chapter 4). Such arenas need to be open, generous, and sharp, driven by curatorship, tools and toolboxes, shared infrastructures, and convening power.

The narrative of the book was filtered through a prism of academia—coming out of academia into the world—and the question of transversal arenas prompted a renewed engagement with *where* capacity building can happen and be enacted. In some ways, universities seem like a bad choice. In any case, what was clear was that capacity of this kind can hardly be singular but rather needs to be networked and distributed. As for nodes or arenas, it would be perfectly possible to turn to civic, activist, communal, foundation-driven, or even corporate contexts and platforms. Efforts could also be state-run (in some countries), low-resource digital ventures, or multi-actor clusters.

Breaking the Gridlock

While some of these models were considered, a basic commitment to the university as a platform for change remained, perhaps not so much driven by the university as it often appears—with heavy disciplinary structures, professional precarity, an inward focus, and a resistance to innovation—but the university that could be: societally engaged sites of learning, ideation, and dialogue; a deeply collaborative and outward-looking place. Two factors were particularly important. Universities can be intermediate actors because they potentially have social and institutional trust and convening power. Also, the humanities (the humanistic, humane) can play an important role in making higher education a stronger societal actor approaching the global challenges of our times.

This argument became clearer through looking at the development over time (chapter 3). Higher education has become a substantial societal sector in the Western world since the massive expansion in the postwar period, eventually turning into an important vehicle for supporting societal progress and innovation, facing a globalized world, and responding to societal challenges. This instrumentalization, heavily entangled with technology, science, and the job market, made institutions of higher education more central but has tended to make the humanities more peripheral. Along with the sector's growth, globalization, and marketization, we saw a shift to a higher degree to corporatization (e.g., corporate-style leadership, PR/communication departments, HR, students as consumers) as well as increased control and monitoring through the expansion of a new public management and "audit culture" model—for example, through economization, harmonization, evaluations, and key performance indicators. In some countries, there has also been an increased politicization of higher education, fueled by the "culture wars." All this has led to more controlled, tightly framed, and risk-averse institutions.

There are at least two ways in which the humanities and the humane (broadly speaking) can contribute to breaking this gridlock. First, an instrumentalized humanities can be a logic-shifting and "edgy" actor in responding collaboratively and responsibly to societal challenges and complex problems, not just strengthening but also altering overall university capacity and direction—in a sense, building infrastructure to make academic and societal infrastructure more humane. Some potential areas and tools were suggested in chapters 3 and 4, but there is obviously no one way.

Second, universities can be unique platforms, building on trust and engagement. This role should be cultivated, strengthened, and scaled up. One way of achieving this is to think of universities as arenas—not for messages from the communication department but rather for vibrant, challenging, and results-driven dialogues. Universities can become transversal arenas driven by ideas, integrity, and careful curation with the ability and will to convene many actors across epistemic, social, and political positions without having to oversell oneself or building new campus institutes. The humanities are well placed to be a driving force in such a development.

Ultimately, the journey toward humane infrastructures is about coming together and being willing to question, push, and challenge boundaries, assumptions, and worldviews—especially when it is hard—to take on problems and challenges that we and our planet face, whether big or small. It is about the instrumentality, capacity building, arenas, tools, but also about the people, joy, fears, and care. It is about curiosity, imagination, and consideration for others (whether human or not), as well as about usefulness, innovation, and results. It is about working together to make the most of our humane and critical capacity to understand, imagine, and build humane infrastructures and a better world.

Notes

Preface

1. Regeringskansliet (Swedish Government), "Nya styrelser för 30 universitet och högskolor."

2. Sveriges universitets- och högskoleförbund (SUHF), "Till regeringen med anledning av beslut om förkortad mandatperiod för universitets- och högskolestyrelserna," April 28, 2023.

3. Wolk, "Sveriges universitetslärare."

4. Cybersecurity and Infrastructure Security Agency CISA (America's Cyber Defense Agency), "Critical Infrastructure Security and Resilience."

5. Collier and Lakoff, "Vital Systems Security," 21–24.

6. Collier and Lakoff, *The Government of Emergency*, 11–12.

7. Dunn Cavelty, "Like a Phoenix from the Ashes," 46–50.

8. For example, for a comparison between the Canadian and US systems, see Boyle and Speed, "From Protection to Coordinated Preparedness."

9. Moteff, Copeland, and Fischer, "Critical Infrastructures," 11.

10. Eriksson and Barck-Holst, "Politik för skydd av kritisk infrastruktur i EU och Sverige—en jämförande analys," 83.

11. Spice, "Fighting Invasive Infrastructures"; Plotnikoff, "Activists or Active Threats?"; Groves, Munday, and Yakovleva, "Fighting the Pipe."

12. Crosby, "The Racialized Logics of Settler Colonial Policing," 422.

13. Reid, "Conclusion: The Biopolitics of Critical Infrastructure Protection," 178.

14. Goldberg, *Dread*; Collier and Lakoff, "The Vulnerability of Vital Systems"; Zebrowski and Sage, "Resilience and Critical Infrastructure."

15. Collier and Lakoff, "The Vulnerability of Vital Systems," 19.

16. European Commission, "European Programme for Critical Infrastructure Protection."

17. Aradau, "Security That Matters," 500.

18. Edwards, "Infrastructure and Modernity."

19. Aradau, "Security That Matters," 492.

20. Sörlin, "The Changing Nature of Environmental Expertise."

21. Spice, "Fighting Invasive Infrastructures."

22. Becker et al., "Pop-up Cycling Infrastructure as a Niche Innovation for Sustainable Transportation in European Cities"; Davies, "Toxic Space and Time."

23. While concepts such as "societal challenges," complex problems," and "missions" are useful, they are not unproblematic. For example, they are often embedded in solutionist frameworks, which tend to privilege certain ways of thinking about goals, actors, and processes.

Introduction

1. Peters, *The Marvelous Clouds*, 30.

2. Svensson, "From Optical Fiber to Conceptual Cyberinfrastructure."

3. Scott-Webber, *In Sync*, 27–30; Karpf, "The Last Dramatic Instructional Innovation?" 78.

4. We also provided chairs for people needing them and would make different arrangements and accommodations in the space as warranted.

5. Mukerji, "Me? A Digital Humanist?" 44–46; Robles-Anderson and Svensson, "One Damn Slide after Another."

6. Hill, "Another Green World."

7. Wikipedia, "List of Public Works Ministries."

8. Sveriges universitets- och högskoleförbund (SUHF), "Dialogseminarium för forskningsfinansiärer och lärosäten (notes from seminar)."

9. Atkins et al., "Revolutionizing Science and Engineering," 5, 12.

10. Mattern, "Interfacing Urban Intelligence"; Peters and Kim, "Critical Humanities Could Bring Good Trouble to AI."

11. Montoya and Leazer, "Limits of 'Infrastructure.'"

12. Liu, "Critical Infrastructure Studies"; Hallinan and Gilmore, "Infrastructural Politics amidst the Coils of Control," 618–619.

13. OECD Global Science Forum, "Report on Roadmapping of Large Research Infrastructures."

14. I owe this comparison to Todd Presner.

15. Collinson, "The Fight to Define Infrastructure Could Change America."

16. Noble, *Algorithms of Oppression*; Miller, "Roads to Nowhere."

17. Robles-Anderson and Svensson, "One Damn Slide after Another."

18. Drucker and Svensson, "The Why and How of Middleware."

19. Estrada, "If You Build It, They Will Move"; Fleischer, "L.A. Freeways Are the Most Racist California Monuments."

20. Strategic Concepts in Organizing and Policy Education (SCOPE), "Between the 110 and the 405."

21. Hughes, *Networks of Power*; Star and Ruhleder, "Steps Toward an Ecology of Infrastructure"; Star, "The Ethnography of Infrastructure"; Bowker and Star, *Sorting Things Out*; Simone, "People as Infrastructure"; Bowker et al., "Toward Information Infrastructure Studies"; Larkin, "The Politics and Poetics of Infrastructure"; Jackson, "Rethinking Repair"; Parks and Starosielski, *Signal Traffic*.

22. D'Ignazio, "What Would Feminist Data Visualization Look Like?"; Brown et al., "Creating Feminist Infrastructure in the Digital Humanities"; Jeremijenko, "Creative Agency and the Space Race of the 21st Century"; Ratto, "Critical Making"; Gilliland and Caswell, "Records and Their Imaginaries"; Drake, *#ArchivesSoWhite in the Words of Jarrett Drake*; McPherson, *Feminist in a Software Lab*; DiSalvo, *Adversarial Design*; Rosner, *Critical Fabulations*; Poirier, "Devious Design."

23. Lundquist, "Här stängslar Gävle Energi bort värmen för en hemlös."

24. Australian Government, "2016 National Research Infrastructure Roadmap," 3.

25. Umeå University, "Humlab."

26. Smarr et al., "A Center for Scientific and Engineering Supercomputing (NSF Proposal)."

27. Henderson, "We the Unhoused."

28. Gordon and Byron, "Sweeping the City," 859.

29. Nguyen, Tweet, April 14, 2020.

30. LA Times Editorial Board, "Editorial."

31. Huang, "Protesters Killed a Plan for a Koreatown Homeless Shelter."

32. Wasserman, "Ralph Ellison, Chester Himes, and the Persistence of Urban Forms."

33. Students4Students, "College Homelessness."

34. UCLA, "Impact | UCLA 100"; "Out of the Classroom, Onto the Streets | UCLA 100."

35. Carroll, Nguyen, and Panaligan, "UCLA Fails to Account for Student Homelessness despite Needing Data to Take Action."

36. It should be noted that there are institutions within UCLA with a very progressive agenda, including the UCLA Luskin Institute on Inequality and Democracy, an important platform that, among other things, organizes an activist-in-residence program, which has included both Nguyen and Henderson.

37. Svensson, "Contemporary and Future Spaces for Media Studies and Digital Humanities," 53.

38. Personal communication with Nick Mitchell (Email. March 9, 2020).

39. Morrison. "Gaps between the Digits."

40. Goh, "Planning the Green New Deal."

41. Risam, *New Digital Worlds*.

42. There is, however, variation over time and between countries. There are also boundary zones, especially between humanities and social sciences.

43. During, "The Idea of the Humanities," 10; Ekström and Sörlin, *Alltings mått*, 7; Hammar and Östling, "Introduction: The Circulation of Knowledge and the History of Humanities," 597. Small, *The Value of the Humanities*, 23.

44. Östling, Jansson, and Stringberg, *Humaniora i välfärdssamhället*; Bod, *A New History of the Humanities*; Hammar and Östling, "Introduction: The Circulation of Knowledge and the History of Humanities"; Small, *The Value of the Humanities*.

45. Braidotti, "A Theoretical Framework for the Critical Posthumanities," 35–36.

46. Frykholm, "Critical Thinking and the Humanities"; Ratto, "Not Just Guns but Bullets, Too."

47. Bardzell and Bardzell, *Humanistic HCI*, 109–110.

48. Kahn, "On Humanities and Human Rights," 117.

49. Hasenöhrl and Meyer, "The Energy Challenge in Historical Perspective"; Rees, "The Many Ways That Water Froze"; Pasek, "Fixing Carbon"; Guédon, "A Curious Halo"; Drucker, "Blind Spot"; Han, "Sonic Pipelines at the Seafloor"; Bullard, "Anatomy of Environmental Racism and the Environmental Justice Movement."

50. It should be noted that many of the key participants in the programming described above are not institutional humanists. However, they tend to have a strong belief in the necessity of such perspectives, which is also demonstrated in their own work.

51. Ekström, "Forging the Integrative Humanities," 285.

1: The Making of Infrastructure

1. Merriam-Webster Online Dictionary, "Infrastructure."

2. Oxford English Dictionary, "Infrastructure, n."

3. Bowker et al., "Toward Information Infrastructure Studies," 97.

4. Fulmer, "What in the World Is Infrastructure?" 32.

5. Larkin, "The Politics and Poetics of Infrastructure," 328.

6. Guajardo Soto, "What Is Infrastructure? Origins, Turns and Continuities of the Concept."

7. Spice, "Fighting Invasive Infrastructures," 40.

8. Collier and Lakoff, "Vital Systems Security," 22–25.

9. Public Safety Canada, "Canada's Critical Infrastructure."

10. Larkin, "The Politics and Poetics of Infrastructure," 328.

11. Burchardt and Höhne, "The Infrastructures of Diversity," 3.

12. Plantin et al., "Infrastructure Studies Meet Platform Studies in the Age of Google and Facebook," 299.

13. Siskin, *System*, 6.

14. Larkin, "The Politics and Poetics of Infrastructure," 39.

15. Los Angeles Department of Transportation, "LADOT Annual Report, January 1, 2022–December 31, 2022," 18.

16. Yu, "Robotaxis Halt Traffic in San Francisco's North Beach Day after Expansion Approval (CBS San Francisco)."

17. Google image search for "infrastructure," January 30, 2017. The first 100 search hits (images) were analyzed.

18. Compare David Nye's notion of the dynamic sublime applied to the introduction and development of the railroad in the US. Nye, *American Technological Sublime*, 45.

19. The first example, infrastructure in the sense of the foundation of a railroad bed (and roads and buildings), is an early use in French.

20. Warde, Robin, and Sörlin, *The Environment*, 12–14.

21. Busch, *Standards*; Hannah and Jeremijenko, "Natalie Jeremijenko's New Experimentalism."

22. Star, "The Ethnography of Infrastructure"; Star and Ruhleder, "Steps Toward an Ecology of Infrastructure."

23. Chiland, "City Officials Give up on Playa Del Rey Road Diets."

24. Svensson, "From Optical Fiber to Conceptual Cyberinfrastructure."

25. Svensson, "From Optical Fiber to Conceptual Cyberinfrastructure."

26. Drucker and Svensson, "The Why and How of Middleware"; Svensson, "From Optical Fiber to Conceptual Cyberinfrastructure."

27. Karasti and Baker, "Infrastructuring for the Long-Term."

28. Svensson, "Humanistisk infrastruktur (main author: Svensson)."

29. European Strategy Forum on Research Infrastructures, "Strategy Report on Research Infrastructures. Roadmap 2016."

30. Directorate-General for Research and Innovation (European Commission), "Research Infrastructures Make Science Happen (Infographic)."

31. Regeringskansliet (Swedish Government), "Stora satsningar på forskning i årets budget."

32. Tillväxtverket, Tweet, November 24, 2017.

33. Ratto and Jackson, "Reopening, Repetition and Resetting," 356:6.

34. Jackson, "Rethinking Repair"; Ramakrishnan, O'Reilly, and Budds, "The Temporal Fragility of Infrastructure."

35. Anand, *Hydraulic City*, 233.

36. Anand, *Hydraulic City*, 187–188.

37. Lee, Dourish, and Mark, "The Human Infrastructure of Cyberinfrastructure," 490–491.

38. Lawhon et al., "Making Heterogeneous Infrastructure Futures in and beyond the Global South," 2.

39. Edwards, "Infrastructure and Modernity," 196.

40. Knödlseder et al., "Estimate of the Carbon Footprint of Astronomical Research Infrastructures"; Walsh, "Your Data Is Dirty"; Gabrys, *Digital Rubbish*; Juckes et al., "Sustainability in Digital Research Infrastructure."

41. Chalmers University of Technology and KTH Royal Institute of Technology, "Climate Framework for Higher Education Institutions—Guidelines"; KTH Royal Institute of Technology, "Klimatramverket."

42. Pasek, "Low-Carbon Research"; Risam and Gil, "Introduction: The Questions of Minimal Computing."

43. Williams, "Infrastructure as Lived Experience (Filmed Keynote Lecture at Landscape Infrastructure Symposium)."

44. Robles-Anderson, "The Crystal Cathedral"; Misharina and Betts, "The Embodied City"; Olofsson, "Gender, Class and Altered Soundscapes."

45. Star and Griesemer, "Institutional Ecology, 'Translations' and Boundary Objects."

46. Bear, *Navigating Austerity*, 180.

47. Bear, *Navigating Austerity*, 204–210.

48. Bear, "'Alternatives' to Austerity," 6.

49. Wright, "Understanding Green Infrastructure," 1013–1014.

50. Efstathiou, "Is It Possible to Give Scientific Solutions to Grand Challenges?"; Kaldewey, "The Grand Challenges Discourse"; Head, *Wicked Problems in Public Policy*.

51. Science|Business, "Lund Declaration"; European Science Foundation, and European Cooperation in Science and Technology, "Responses to Environmental and Societal Challenges for Our Unstable Earth (RESCUE)."

52. Pratt, "Arts of the Contact Zone."

53. UCLA Humanities Division, "UCLA Experimental Humanities Workshop on 'Humane Infrastructures.'"

54. Peters and Turner, "Unlock the Volume."

55. Rothstein, "Myndigheter att lita på"; Sörlin, "Så blev Stockholm en världsstad i kampen mot miljöförstörelsen."

56. Aselmeyer, "Capturing the Zeitgeist of Infrastructure."

57. *The New York Times*, July 2, 1950; *Springfield Daily News*, March 1, 1952; *Dagens Nyheter*, December 29, 1952; the *Frankfurter Allgemeine Zeitung* March 1, 1952.

58. Laak, *Lifelines of Our Society*, 7.

59. Wright, "Understanding Green Infrastructure," 1013.

60. Svensson, *Number and Countability in English Nouns*, 153–158.

61. Countability is more likely when talking about specific types of infrastructure, such as "research infrastructures," or more generally in the context of counting infrastructures projects being given funding or placed on a road map.

62. Samorani et al., "Overbooked and Overlooked."

63. Larkin, "The Politics and Poetics of Infrastructure," 330.

64. Borgman, *Scholarship in the Digital Age*, 22.

65. Carse, "Keyword: Infrastructure," 28–29.

66. Williams, *Keywords*, 15.

67. Star and Griesemer, "Institutional Ecology, 'Translations' and Boundary Objects," 393.

68. NATO, "Aspects of NATO Series 1—No. 12. Infrastructure and Logistics—NATO Archives Online."

69. North Atlantic Treaty Organization, Infrastructure Committee, *50 Years of Infrastructure*, 19.

70. North Atlantic Treaty Organization, Infrastructure Committee, *50 Years of Infrastructure*, 17.

71. Aagaard Jensen, "Connecting the Alliance," 192–194.

72. Aagaard Jensen, "Connecting the Alliance," 209–210.

73. Kennedy, "John F. Kennedy: Remarks of Welcome to President Ayub Khan of Pakistan at Quonset Point, Rhode Island."

74. UC Santa Barbara, "The American Presidency Project." The search results were retrieved December 19, 2023 through a document archive search for "infrastructure," specifying all John F. Kennedy materials.

75. The data from the American Presidency Project (John F. Kennedy–Barack Obama) have been adjusted for the number of "presidential documents" for each president, but there are still several uncertainties, including possible variation for document size across presidencies. The quotient Clinton–Reagan in the data is 3.8 and the ratio George W. Bush–Reagan is 5.9. The quotient Obama–his two predecessors (together) in the data is 1.7.

76. Donald J. Trump website, "Infrastructure (Donald J. Trump's Vision)."

77. Altshuler, "Review: Infrastructure Investment," 506.

78. An example: The first citation with "infrastructure" listed by the *Oxford English Dictionary* from 1927 is sometimes taken as evidence of an active general use of the term from the late 1920s. It is a rare occurrence, however (the next citation in the OED article is from 1950) and, moreover, from a description of a French railway, and—as the

actual underlying magazine article shows—it is juxtaposed with "superstructure." This use is very clearly influenced by the French usage and context. See also Carse, "Keyword: Infrastructure," 27.

79. Carse, "Keyword: Infrastructure," 33.

80. Kungliga biblioteket, "Svenska Dagstidningar (Swedish Newspapers)."

81. A simple comparison can be made through searching for the word "teknik" ("technology" or "technique") during the same time range. Result: 41,591 newspaper pages.

82. All searches were made using the web portal for "Svenska dagstidningar," collecting the number of newspaper pages containing "infrastruktur" (as a whole word) for *Dagens Nyheter*, *Svenska Dagbladet*, *Aftonbladet*, and *Expressen* for each decade. The data was originally collected in June 2018 and updated in August 2024 (with very minor changes, probably because a new version of the service had been released).

83. The official governmental data service was used to identify documents containing the search term "infrastruktur." The data were verified in August 2024. Riksdagsförvaltningen (Swedish Government), "Dokument & lagar." The relative difference between 1984 and 1991 was calculated based on the number of instances and total number of documents for the years in question (1984: 2,984 documents, 25 instances; 1991: 4,857 documents, 281 instances). The overall pattern was observed through using an auxiliary service that maps absolute and relative frequencies over time in the material in question. GRIDH, University of Gothenburg, Chalmers University of Technology, and Riksdagens öppna data, Swedish Government. "Hitta och läs riksdagsmotioner."

84. Ohlin, Motion 1961:363 Andra kammaren—Riksdagen.

85. Riksdagsförvaltningen (Swedish Government), "Dokument & lagar."

86. Dempsey, "The Rise and Fall of the Interstate Commerce Commission."

87. Bear, "'Alternatives' to Austerity," 6.

88. Steinhart and Kittay, "Generating Metaphors from Networks," 41.

89. Davies, "Corpus of Historical American English (COHA)." The data were collected in December 2020.

90. There is one instance in the corpus of "public works infrastructure" that was also used for the title of a US government report from 1983 (CBO).

91. COHA contains more than 475 million words of text from the 1820s–2010s. It is a balanced or structured historical corpus of considerable size, which makes it possible to compare materials across genres and decades with a good degree of granularity and robustness.

92. Choate and Walter, *America in Ruins*.

93. Choate and Walter, *America in Ruins*.

94. The sample used includes up to page 30, excluding tables, figures, and notes but including the introduction. Acronyms and abbreviations (such as "LPW" for "Local Public Works") were excluded. All instances were counted including noun adjuncts

(such as "public works investments") and names of departments or programs. Five of six instances of "infrastructure/s" were preceded by "public."

95. Patton, "Infrastructure Decay in the United States"; Berend, "Development of the Infrastructure—in Historical Perspective"; Congressional Budget Office, Congress of the United States, "Public Works Infrastructure"; National Council on Public Works Improvement, *Fragile Foundations*.

96. Hanke, "Shaky Foundation, Overbuilt Solution."

97. The first report card was published in 1988 by the National Council on Public Works Improvement (grade C), which was followed by ASCE reports from 1998.

98. American Society of Civil Engineers, "ASCE's 2017 Infrastructure Report Card."

99. Zeller, "Aiming for Control, Haunted by Its Failure," 209.

100. Vetenskapsrådet, "Årsredovisning 2001."

101. Vetenskapsrådet, "Årsredovisning 2022."

102. Vetenskapsrådet, "Årsredovisning 2022," 78–79.

103. Bengtsson et al., "Forskningen hotas."

104. European Commission, "Horizon Europe"; National Science Foundation, "FY 2024 Budget Request to Congress (NSF)," 9.

105. Bell and Dourish, "Yesterday's Tomorrows."

106. Atkins et al., "Revolutionizing Science and Engineering," 31.

107. American Council of Learned Societies (ACLS), "Our Cultural Commonwealth," 40.

108. Borgman, *Scholarship in the Digital Age*, 30.

109. Government of Canada, Budget Implementation Act, 1997; Association of Universities and Colleges of Canada, Canadian Association of University Teachers, and National Consortium of Scientific and Educational Societies, "Putting Knowledge to Work."

110. Canada Foundation for Innovation, "Our Mandate."

111. Hallonsten, "Research Infrastructures in Europe," 630.

112. Canada Foundation for Innovation, "Our Mandate."

113. The European Spallation Source, "About ESS."

114. Canada Foundation for Innovation, "Our Mandate."

115. Lopreite and Murphy, "The Canada Foundation for Innovation as Patron and Regulator," 124.

116. Lopreite and Murphy, "The Canada Foundation for Innovation as Patron and Regulator," 124.

117. Lopreite and Murphy, "The Canada Foundation for Innovation as Patron and Regulator," 138.

118. Morgan, "Higher Education Funding and Policy Trade-Offs," 70–71.

119. There are other academic infrastructures, however, that have more time depth. The National Library in Sweden for example, which has its origins in the sixteenth century, does not have a specific time extension.

120. European Commission, "Statues of the European Spallation Source ERIC," 23.

121. European Commission, "Statues of the European Spallation Source ERIC," 26.

122. European Parliament and the European Council, "Decision No 182/1999/EC of the European Parliament and of the Council of 22 December 1998 concerning the fifth framework programme of the European Community for research, technological development and demonstration activities (1998 to 2002)."

123. Papon, "European Scientific Cooperation and Research Infrastructures," 72.

124. European Commission, "Towards a European Research Area," 10–11.

125. European Strategy Forum on Research Infrastructures, "European Roadmap for Research Infrastructures. Report 2006," 10.

126. Cramer et al., "Big Science and Research Infrastructures in Europe," 5–6.

127. Zabusky, *Launching Europe*, 50.

128. Bolliger and Griffiths, "The Introduction of ESFRI and the Rise of National Research Infrastructure Roadmaps in Europe," 120.

129. Bolliger and Griffiths, "The Introduction of ESFRI and the Rise of National Research Infrastructure Roadmaps in Europe," 122.

130. Benner, "Big Science in a Small Country," 162.

131. Bolliger and Griffiths, "The Introduction of ESFRI and the Rise of National Research Infrastructure Roadmaps in Europe," 102; Phaal, Farrukh, and Probert, "Technology Roadmapping—A Planning Framework for Evolution and Revolution."

132. Phaal, Farrukh, and Probert, "Technology Roadmapping—A Planning Framework for Evolution and Revolution," 23.

133. Womersley, "More Effective Coordination of National Roadmaps."

134. European Strategy Forum on Research Infrastructures, "European Roadmap for Research Infrastructures. Report 2006," 7.

135. OECD Global Science Forum, "Report on Roadmapping of Large Research Infrastructures," 6.

136. Sveriges unga akademi and Sällskapet riksdagsledamöter och forskare (RIFO), "Framtiden för svensk forskningsinfrastruktur (seminar)."

137. European Strategy Forum on Research Infrastructures, Long-Term Sustainability Working Group, "Long-Term Sustainability of Research Infrastructures," 66.

138. It should be noted, however, that there can be several parallel trajectories. In the mid to late 1990s, Riksbankens Jubileumsfond (RJ)—a major Swedish foundation for the humanities and social sciences—initiated funding schemes for infrastructure

projects to meet the demands from cultural heritage institutions, mainly for digitization efforts (Riksbankens Jubileumsfond, "Redogörelse till Riksdagen 1996/97:RJ").

139. Data retrieved from the Wayback Machine on November 17, 2019. The category was introduced in the following date frames: Stockholm University (October 20–December 3, 2014), Gothenburg University (February 6–March 9, 2016), Uppsala University (November 21, 2016–January 18, 2017, with a special focus on research infrastructure, which changed into a full category toward the end of the year). At the time of the data excerption, Linköping University and Lund University did not have categories for "research infrastructure" under "research," and seemingly nowhere else at university level (based on searches). Additional data regarding Linköping and Lund were retrieved on December 28, 2023. The change reported for Lund University was effectuated on February 25, 2023, but was not in effect on February 19, 2023, according to Wayback Machine data.

140. Uppsala University, "Infrastruktur."

141. Stockholm University, "Forskningsinfrastruktur."

142. Umeå University, "Forskningsinfrastruktur."

143. Hansson, "Yttrande från Kungl. Vetenskapsakademien inför den kommande forskningspolitiska propositionen 2016."

144. URFI (Universitetens referensgrupp för forskningsinfrastruktur) was established in 2010 through the Principals' Council associated with the Knut and Alice Wallenberg Foundation. See Riklund, "Del 1:2 Forskningsinfrastruktur," Umeå University, "Universitetsledningens externa uppdrag," and Bengtsson et al., "Forskningen hotas."

145. Riklund, "Del 1:2 Forskningsinfrastruktur."

146. Riklund, "Del 1:2 Forskningsinfrastruktur."

147. Lopreite and Murphy, "The Canada Foundation for Innovation as Patron and Regulator," 137.

148. European Strategy Forum on Research Infrastructures, "ESFRI Roadmap 2021, Projects and Landmarks."

149. Sumathipala, "Research Infrastructure," 409, 411; European Commission and Canada Foundation for Innovation, "International Conference on Research Infrastructures Proceedings Report June 1–3, 2021."

150. The International Union of Pure and Applied Physics, "UL Opara: 'With Research Infrastructure, Young Africans Will Perform Wonders.'"

151. European Commission, "Horizon Europe Work Programme 2023–2024, 3. Research Infrastructures (European Commission Decision C(2024) 2371 of 17 April 2024)," 10.

152. Vetenskapsrådet, "Vetenskapsrådets guide till infrastrukturen 2014"; GBIF, "Global Biodiversity Information Facility. Free and Open Access to Biodiversity Data."

153. The covers show scientific tools of measurement—a compass and dividers on top of a map—or a cartographic network visualization based on species occurrence data.

154. CLARIN, "CLARIN ERIC"; DARIAH-EU, "DARIAH | Digital Research Infrastructure for the Arts and Humanities."

155. European Commission, Directorate-General for Research and Innovation, "Supporting the Transformative Impact of Research Infrastructures on European Research," 154.

156. ESFRI Working Group on Monitoring of Research Infrastructures Performance, "Monitoring of Research Infrastructures Performance."

157. ESFRI Working Group on Monitoring of Research Infrastructures Performance, "Monitoring of Research Infrastructures Performance," 4.

158. Dalingwater, "Post-New Public Management (NPM) and the Reconfiguration of Health Services in England"; Edwards, "Infrastructure and Modernity."

159. Vetenskapsrådet, "The Swedish Research Council's Guide to Research Infrastructures 2012," 9.

160. Runting, Sjögrim, and Torisson, "Pop Theory," 517–519.

161. Hallonsten, "Introduction. In Pursuit of a Promise," 12.

162. Granberg, "The ESS Project as a Generator of Conflict and Collaboration"; Kungl. Vetenskapsakademien (Royal Swedish Academy of Sciences), "Forskningspropositionen 2008/09:50 rörande MAX IV och ESS (letter to the Government)."

163. Marx, *The Machine in the Garden*, 150; Vickery, *Landscape and Infrastructure*, 21.

164. Nye, *American Technological Sublime*, xiii.

165. Granberg, "The ESS Project as a Generator of Conflict and Collaboration," 203.

166. Australian Government, "2016 National Research Infrastructure Roadmap," 1.

167. Mukerji, *Impossible Engineering*, 16.

168. D'Ignazio and Klein, *Data Feminism*; Svensson, "Critical Visualization Event at the Graduate Center, CUNY on June 9, 2016"; Drucker and Svensson, "The Why and How of Middleware."

169. Vetenskapsrådet, "Forskningsinfrastrukturdagen 2018."

170. Vetenskapsrådet, "Forskningsinfrastrukturdagen 2018 (Video)."

171. I attended the principal session in person and have watched the recording from the whole day. It should be noted that there was one female speaker in the afternoon program.

172. World Trade Center Stockholm, "Conference Facilities | WTC."

173. These and other data and observations come from the program and from attending the event. I do not discuss ethnicity or race here, but it is worth noting that all these men were white.

174. Vetenskapsrådet, "Vetenskapsrådets årsredovisning 2016," 20.

175. My translation.

176. Vetenskapsrådet, "Research Infrastructure."

177. Four women applicants, but three individuals, as one person had two successful applications.

178. Vetenskapsrådet, "Natural and Engineering Sciences." There were seventy approved applications for women (21.5%) and 255 for men (78.5%). It can be noted that the decision from the council of Natural and Engineering Sciences contains gender-based statistics whereas there is no equivalent information in the infrastructure decision.

179. Smith, "The Human Touch Software of the Highest Order," 4.

180. Koh and Risam, "Open Thread."

181. Svensson, "Humanities Computing as Digital Humanities"; Svensson, *Big Digital Humanities*.

182. McPherson, "Theory/Practice," 14.

183. Infrastructure and Projects Authority, UK, "National Infrastructure Delivery Plan 2016 to 2021," 5.

184. American Council of Learned Societies (ACLS), "Our Cultural Commonwealth"; European Science Foundation, "Research Infrastructures in the Digital Humanities. Science Policy Briefing"; All European Academies (ALLEA), "Going Digital."

185. Australian Government, "2016 National Research Infrastructure Roadmap," 36.

186. Vetenskapsrådet, "Research Overview 2019," 17.

187. Chalmers University of Technology, "Research Infrastructure at Chalmers."

188. Chalmers University of Technology, "Research Infrastructure at Chalmers."

189. Chalmers University of Technology, "GENIE Gender Report 2018."

190. Chalmers University of Technology, "A Big Investment to Make Chalmers Equal."

191. Chalmers University of Technology, "GENIE Gender Report 2018"; Chalmers University of Technology, "GENIE Gender Report 2019."

192. Palm, "Storsatsning på jämställdhet gav 'inga förändringar.'"

193. Costanza-Chock, *Design Justice*; Ratto, "Critical Making."

194. Gebru et al., "Statement from the Listed Authors of Stochastic Parrots on the 'AI Pause' Letter."

195. Amoore, "Machine Learning Political Orders," 25.

2: The Infrastructural Humanities

1. Scott-Webber, *In Sync*.

2. University of Warwick, "Arts Faculty Facilities"; Eby, "Humanities Hall Ready to Welcome Students"; Lee, "New State-of-the-Art Humanities Building to Replace Susan B. Davis Residence Hall, Former Zoe Bayliss Co-Op"; Huynh, "UW–Madison Building Projects."

3. Ridder-Symoens, *A History of the University in Europe*, 33–41; Alemu, "Meaning, Idea and History of University/Higher Education in Africa," 210–214.

4. Banfill, Presner, and Zubiaurre, "Urban Humanities Pedagogy," 120.

5. Coulson, Roberts, and Taylor, *University Planning and Architecture*.

6. Hansson, "Konsten att vara vetenskaplig," 25; Clark, "On the Dialectical Origins of the Research Seminar," 119–120; Risbjerg Eskildsen, "Virtues of History," 29–30.

7. Palmieri, "The Forgotten Seminar," 77.

8. Smith, "Gender and the Practices of Scientific History," 1157–1159.

9. Risbjerg Eskildsen, "The Humanities in the Vocational University," 271.

10. Karlsohn, "The Academic Seminar as Emotional Community."

11. Smith, "Gender and the Practices of Scientific History," 1160.

12. Hughes, *Networks of Power*; van der Vleuten, "Large Technical Systems."

13. Star and Ruhleder, "Steps Toward an Ecology of Infrastructure"; Star, "The Ethnography of Infrastructure"; Edwards et al., "Understanding Infrastructure"; Bowker et al., "Toward Information Infrastructure Studies."

14. Fidler and Russell, "Financial and Administrative Infrastructure for the Early Internet"; Henke and Sims, *Repairing Infrastructures*; Plantin and Punathambekar, "Digital Media Infrastructures"; Aradau, "Security That Matters."

15. Noble, *Algorithms of Oppression*; Bear, *Navigating Austerity*; Posner, "Breakpoints and Black Boxes"; Gabrys, "Programming Nature as Infrastructure in the Smart Forest City"; Thylstrup, *The Politics of Mass Digitization*.

16. Venkatesan et al., "Attention to Infrastructure Offers a Welcome Reconfiguration of Anthropological Approaches to the Political," 4.

17. Society for Cinema and Media Studies. "Past SCMS Conferences."

18. These data were calculated from the conference booklets available on the SCMS website, including the booklet for 2020 (which ended up being canceled). All instances of "infrastructure" in session and paper titles were registered (advertisements, prizes etc. were disregarded). The numbers given are absolute, and the conference has grown over the years (the 2006 booklet contains 94 pages whereas the 2020 booklet includes 244 pages), but regardless, the development indicated must be considered certain.

19. Parks and Starosielski, *Signal Traffic*; Plantin and Punathambekar, "Digital Media Infrastructures," 165–166.

20. Johnson and Zieger, "Reading for Infrastructure, Infrastructure for Reading (UCHRI Short-Term Collaborative Residency)."

21. Johnson and Zieger, "Reading for Infrastructure, Infrastructure for Reading (UCHRI Short-Term Collaborative Residency)."

22. Arbona et al., "Reading for Infrastructure."

23. Sandvig, "The Internet as Infrastructure," 89.

24. Parks and Starosielski, "Introduction."

25. Parks, "Stuff You Can Kick," 357.

26. Parks, "Stuff You Can Kick," 370.

27. Liu, "Critical Infrastructure Studies."

28. Peters, *The Marvelous Clouds*, 31.

29. Peters, *The Marvelous Clouds*, 33.

30. The Daily Californian, "UCPath Has Left Employees Facing Financial Instability (Daily Californian Editorial)."

31. Svensson, "Universitetet som uttryck," 99.

32. Smith, "The Human Touch Software of the Highest Order"; Robles-Anderson and Svensson, "One Damn Slide after Another"; Svensson, "Contemporary and Future Spaces for Media Studies and Digital Humanities"; Pasek, "Low-Carbon Research."

33. Livingstone, *Putting Science in Its Place*; Oman, *Higher Education and the Gendering of Space in England and Wales, 1869–1909*.

34. Kaufmann-Buhler, *Open Plan*, 16.

35. Gärdebo et al., "Klimatforskarna förvärrar utsläpp."

36. Eriksson et al., "On the Necessity of Flying and of Not Flying"; Wormbs, "Därför måste vi våga moralisera i klimatfrågan."

37. McLean, "Jane Ohlmeyer Is an Academic Unafraid of Trinity's Political Tightrope."

38. European Strategy Forum on Research Infrastructures, "Strategy Report on Research Infrastructures: Roadmap 2016," 14–15. Both projects and landmarks were analyzed based on their classification into different areas and the available cost specification. If there was a range indicated, the lower value was used. One source of error was the unavailability of values for some entries. These are approximate, "rough" data. In any case, the pattern—one of scale—is clear for landmarks as well as projects. For projects (only one categorized as social and cultural innovation), the share of construction costs was 0.2% and 1.8% for annual running costs.

39. In an earlier and more provisional account, I proposed a categorization of humanities infrastructure based on three positionalities: basing "infrastructure" on what is already there, the conception that no infrastructure is needed, and adopting a technology- and science-driven model. See Svensson, *Big Digital Humanities*, 144–152.

40. Ayers, "Where the Humanities Live," 29.

41. Brontë, *A Guide to Humanities Centers in the United States*, 244.

42. Baca, "Getty Voices."

43. Holm, Jarrick, and Scott, *Humanities World Report 2015*, 149–150.

44. Paoletti, Corbin Sies, and Jenkins, "Using a Virtual Museum for Collaborative Teaching, Research, and Service," 151; UC Riverside, "2nd Annual Digital Humanities Infrastructure Symposium."

45. Matres, Oiva, and Tolonen, "In Between Research Cultures," 54.

46. European Strategy Forum on Research Infrastructures, "Roadmap for European Research Infrastructures," 8.

47. Yates and Murphy, *Engineering Rules.*

48. Peters, *The Marvelous Clouds*, 28.

49. Thornton, *The Scholar in His Study*, 127.

50. The National Gallery, London, "Building the Picture | Antonello Da Messina | Saint Jerome in His Study."

51. Facebook conversation initiated on May 7, 2017, by Christer Nordlund.

52. Ehn and Löfgren, *Hur blir man klok på universitetet?* 69.

53. Ehn and Löfgren, *Hur blir man klok på universitetet?* 70.

54. Smith, "Gender and the Practices of Scientific History," 1174–1175.

55. HASS Committee on the Humanities Library, MIT, "Final Report"; Svensson, "From Optical Fiber to Conceptual Cyberinfrastructure."

56. HASS Committee on the Humanities Library, MIT, "Final Report."

57. European Science Foundation, "Research Infrastructures in the Digital Humanities. Science Policy Briefing," 3.

58. Mattern, "Library as Infrastructure."

59. Norberg, *Mellan tiden och evigheten*; Nordin, "Plikten bakom allt."

60. Svensson, "Kapitel 4: Nationalbiblioteket i Humlegården, 1800-talet," 165–166.

61. Sveriges riksdag, Förordning (2008:1421) med instruktion för Kungl. biblioteket (fram till SFS 2018:15).

62. American Council of Learned Societies (ACLS), "Our Cultural Commonwealth," 11.

63. American Council of Learned Societies (ACLS), "Our Cultural Commonwealth," 18.

64. Head, "Envisioning Biblotheca 2.0."

65. European Science Foundation, "Research Infrastructures in the Digital Humanities. Science Policy Briefing," 2.

66. Amréus et al., "Digitalisera kulturarvet för demokratins skull" (DN Debatt).

67. Thylstrup, *The Politics of Mass Digitization*, 107–108.

68. Dahlström, Hansson, and Kjellman, "'As We May Digitize'—Institutions and Documents Reconfigured," 461.

69. Klein, "Image of Absence," 665.

70. Svensson, Sörlin, and Wormbs, "The Movement Heritage."

71. Snickars, *Kulturarvets mediehistoria*, 42.

72. American Council of Learned Societies (ACLS), "Our Cultural Commonwealth," 6.

73. Caswell and Mallick, "Against Precarity."

74. Berg et al., "Making Archival Speech Recordings Accessible for Research."

75. Brown and Nicholas, "Protecting Indigenous Cultural Property in the Age of Digital Democracy," 309, 318–319.

76. Drucker, *Visualization and Interpretation*, 5.

77. Drucker and Svensson, "The Why and How of Middleware."

78. Presner, "The Ethics of the Algorithm," 191–192.

79. Presner, "The Ethics of the Algorithm," 201.

80. School of Education and Information Studies, UCLA, "Johanna Drucker: The Future of Special Collections Libraries."

81. Kungliga biblioteket, "Årsredovisning 2019"; Riksarkivet, "Årsredovisning 2019"; Kungliga biblioteket, "Årsredovisning 2020"; Riksarkivet, "Årsredovisning 2020."

82. Kungliga biblioteket, "Årsredovisning 2021"; Riksarkivet, "Årsredovisning 2021"; Kungliga biblioteket, "Årsredovisning 2022"; Riksarkivet, "Årsredovisning 2022."

83. Kungliga biblioteket, "Färdplan 2023–2025."

84. Corresponding Swedish words used in the analysis of the annual reports: "hållbarhet," "miljö," "ekologi," and "klimat."

85. Social Sciences and Humanities Research Council, "Community Consultation on Infrastructure in the Social Sciences and Humanities," 3.

86. Social Sciences and Humanities Research Council, "Community Consultation on Infrastructure in the Social Sciences and Humanities," 3.

87. Vetenskapsrådet, "Research Overview 2019," 16.

88. Edwards, *A Vast Machine*, 187–188; Loukissas, *All Data Are Local*, 37–39; Pelizza, "Developing the Vectorial Glance," 305.

89. Bhattacharyya. "Epistemically Produced Invisibility."

90. Pelizza, "Developing the Vectorial Glance," 305.

91. Vetenskapsrådet, "Research Overview 2019," 18.

92. European Strategy Forum on Research Infrastructures, "Social Sciences and Humanities: Roadmap Working Group Report 2008," 7.

93. Atkinson et al., "Data-Driven Research in the Humanities—the DARIAH Research Infrastructure."

94. National Endowment for the Humanities, "Digging into Data Challenge"; Vetenskapsrådet, "DIGARV—Om DIGARV."

95. European Strategy Forum on Research Infrastructures, "Roadmap 2018: Strategy Report on Research Infrastructures."

96. Atkins et al., "Revolutionizing Science and Engineering."

97. American Council of Learned Societies (ACLS), "Our Cultural Commonwealth," 8.

98. American Council of Learned Societies (ACLS), "Our Cultural Commonwealth," 8.

99. Svensson, *Big Digital Humanities*, 150–151.

100. Smithies, *The Digital Humanities and the Digital Modern*, 58.

101. European Open Science Cloud (EOSC), "About Page, EOSC."

102. European Commission, "European Open Science Cloud."

103. European Open Science Cloud (EOSC), "TEXTCROWD (EOSCPilot)."

104. European Open Science Cloud (EOSC), "EOSCpilot Demonstrator: TEXTCROWD."

105. Edwards, *A Vast Machine*.

106. Riber and Jackson, "Data Bite Man"; Loukissas, *All Data Are Local*.

107. Bowker, "Data Flakes."

108. Noble, *Algorithms of Oppression*; D'Ignazio and Klein, *Data Feminism*.

109. Ratto, "Epistemic Commitments, Virtual Reality, and Archaeological Representation."

110. Australian Government, "2016 National Research Infrastructure Roadmap," 3.

111. Government Bill (House of Commons) C-93 (35-2)—Royal Assent—An Act to implement certain provisions of the budget tabled in Parliament on February 18, 1997—Parliament of Canada.

112. Hammar and Östling, "Introduction: The Circulation of Knowledge and the History of Humanities."

113. Modern Language Association, "The Future of Scholarly Publishing: MLA Ad Hoc Committee on the Future of Scholarly Publishing"; European Commission, Directorate General for Research and Innovation, *Future of Scholarly Publishing and Scholarly Communication*; Davidson, "The Futures of Scholarly Publishing."

114. Ågren and Engdahl, "Gör konsekvensanalys innan tidskriftsstödet avskaffas."

115. Watkinson and Pitts, "Re-Envisioning Humanities Infrastructure."

116. Watkinson and Pitts, "Re-Envisioning Humanities Infrastructure."

117. Fitzpatrick, *Planned Obsolescence*, 7.

118. Deb, "Trump Proposes Eliminating the Arts and Humanities Endowments."

119. National Endowment for the Humanities, "NEH Chairman William D. Adams Statement on the Proposed Elimination of NEH in FY18 Budget | National Endowment for the Humanities."

120. Leef, "The Case for Abolishing the National Endowment for the Humanities Just Got Stronger (Opinion Piece)."

121. National Endowment for the Humanities, "Strategic Plan, Fiscal Year 2013–Fiscal Year 2017"; National Endowment for the Humanities, "About the National Endowment for the Humanities."

122. Woodward, "The Future of the Humanities—in the Present and in Public"; Ellison, "Guest Column—The New Public Humanists," 290.

123. Woodward, "The Future of the Humanities—in the Present and in Public," 116–117.

124. Mullen, "Public Humanities' (Victorian) Culture Problem," 186–192.

125. Chandler, "Vulnerability Without and Within."

126. Liddle, "The Rise of Public Scholarship."

127. Östling et al., "Kunskapens nya rörelser."

128. Östling et al., "Kunskapens nya rörelser," 4, 24, 27, 31.

129. Östling et al., "Kunskapens nya rörelser," 40.

130. Drucker and Svensson, "The Why and How of Middleware."

131. Östling et al., "Kunskapens nya rörelser," 17–18.

132. Ivković and Trajković, "Public Humanities Today," 110–111.

133. Mullen, "Public Humanities' (Victorian) Culture Problem," 197.

134. Svensson, *Big Digital Humanities*.

135. UC Humanities Research Institute and San Diego Supercomputer Center at UC San Diego, "Summer Institute: Cyberinfrastructure; Svensson, *Big Digital Humanities*, 8–9."

136. Interview with Donald Robotham by Patrik Svensson, May 18, 2016. Unpublished recording.

137. Padberg, "The Global Diversity of Institutes for Advanced Study," 130–133.

138. The Institute for Advanced Study Princeton, "The Institute for Advanced Study."

139. I draw partly on collaborative work with David Theo Goldberg on humanities centers.

140. Ekström and Sörlin, "The Integrative Humanities—and the Third Research Policy Regime," 202–203.

141. University of Wisconsin–Madison. "History (Institute for Research in the Humanities)". The Center for the Humanities at Wesleyan University is sometimes also listed for the same year (it was the Center for Advanced Study until 1969), see Wesleyan University, "About the Center, Center for the Humanities—Wesleyan University."

142. Cornell University. "About Us | Society for the Humanities"; McCabe, "Hopkins Humanities Center Celebrates 50 Years as Home to a Diverse Intellectual Community."

143. I owe the model of three waves to David Theo Goldberg (personal communication).

144. Barclay and Turner, *Humanities Research Centre*, 1, 6.

145. There are some typologies suggested in the literature. For example, Brontë's four categories in the Rockefeller Foundation Guide (traditional research centers part of a university, traditional research centers free of any university connections, nontraditional research centers, and centers focused on public program activity) and Kaplan's division of institutes into eclectic (multidisciplinary) and integrative (often theory focused) along one axis and separate (the science model, a separate intellectual space devoted to cutting-edge research) and connected (part of the daily life of the institution, more open) along the other axis; Kaplan, "What Impact Should, or Can a Humanities Institute Have on the Organization of Knowledge in the University?"

146. Brontë, *A Guide to Humanities Centers in the United States.*

147. Wittrock, "A Brief History of Institutes for Advanced Study."

148. Padberg, "The Global Diversity of Institutes for Advanced Study," 134.

149. Padberg, "The Global Diversity of Institutes for Advanced Study," 157.

150. Stanford Humanities Center, "2015–16 Annual Report: Dialogue."

151. Giamatti, "Introductory Remarks," 2.

152. Institute for Advanced Study Princeton, "IAS Welcomes Over 250 Scholars for 2022–23 Academic Year—Press Release | Institute for Advanced Study."

153. Brontë, *A Guide to Humanities Centers in the United States*, 254.

154. Donaldson, "The Humanities Research Centre, The Australian National University," 4.

155. Based on six Stanford Humanities Center annual reports: 2011–2012, 2012–2013, 2013–2014, 2014–2015, 2015–2016, and 2016–2017. Publications are not included as a category in more recent annual reports (2017–2020).

156. University of Wisconsin–Madison, "Institute for Research in the Humanities."

157. National Humanities Center, "Fellows by the Numbers."

158. Walsh, "Behaviorial Sciences," 655.

159. Swedish Collegium for Advanced Study, "SCAS: The Natural Sciences Programme."

160. Schnapp and Shanks, "Artereality (Rethinking Craft in a Knowledge Economy)"; Winn, "Interpretation as Performance."

161. Woodward, "Power and Place," 2.

162. Carnochan, "Institutes and the Conscience of the King," 4.

163. Including David Theo Goldberg (the UC Humanities Research Institute), Kathleen Woodward (the Simpson Center for the Humanities), and James Chandler (the Franke Institute).

164. Goodheart, "Reflections on the Institutional Impact of Institutes," 2.

165. Wesleyan University, "About the Center, Center for the Humanities—Wesleyan University."

166. The Franke Institute for the Humanities, the University of Chicago, "The Franke Institute for the Humanities."

167. Woodward, "Power and Place," 2.

168. Besedovsky et al., "Time as Infrastructure."

169. MacCannell, "Institutes and Institutions," 3-4.

170. European Commission, "About Research Infrastructures | Infrastructures—Research and Innovation."

171. Svensson, "The Humanistiscope."

172. European Spallation Source, "European Spallation Source."

173. European Spallation Source, "European Spallation Source."

3: Instrumentalizing the Humanistic

1. Heise, "Terraforming for Urbanists"; McPherson, "Designing for Difference"; Juul Søndergaard, *Staying with the Trouble through Design*; Bowker, *Science on the Run*.

2. Heise, "Terraforming for Urbanists," 23.

3. Heise, "Terraforming for Urbanists," 19.

4. Masco, "The Age of Fallout."

5. Masco, "The Age of Fallout," 164.

6. Bratton, "The Terraforming"; Strelka Institute/Институт Стрелка and Bratton, "The Terraforming 2022. New Program Presentation (Video)."

7. Nye, *American Technological Sublime*.

8. Bratton, "Planetary Sapience."

9. Nye, *American Technological Sublime*, xiii.

10. Harvard School of Engineering and Applied Sciences, "Model Shows Solar Geoengineering May Be Surprisingly Effective in Alleviating Impacts of Global Warming on Crops."

11. Ratto, "Making at the End of Nature."

12. Arizona State University, "PlanetWorks—Designing and Managing the Future of Planet Earth"; Stanford University, "Introducing the Stanford Doerr School of Sustainability."

13. Stanford University, "Introducing the Stanford Doerr School of Sustainability"; Arizona State University, "Like a Medical Center for the Planet."

14. Szeman and Boyer, "Introduction: On the Energy Humanities," 9.

15. SHAPE-ID: Shaping Interdisciplinary Practices in Europe, "Christian Pohl Discusses the Role of the Humanities Addressing Societal Challenges (Video Recording)."

16. Ekström and Sörlin, *Alltings mått*, 220.

17. Bowker and Star, *Sorting Things Out*, 34.

18. Bowker and Star, *Sorting Things Out*, 35.

19. Bowker, *Science on the Run*, 13–14.

20. Bowker, *Science on the Run*, 167.

21. Edwards, *A Vast Machine*, 20.

22. Edwards, *A Vast Machine*, 20–21.

23. Dumit, "Writing the Implosion"; Noble and Roberts, "Out of the Black Box"; Ratto, "Epistemic Commitments, Virtual Reality, and Archaeological Representation."

24. Qadri, Tweet, July 12, 2021.

25. Dumit, "Writing the Implosion."

26. Horton, *The Cosmic Zoom*; Powell, *Undoing Optimization*.

27. D'Ignazio and Klein, *Data Feminism*; Loukissas, *All Data Are Local*; Kitchin, *The Data Revolution*.

28. Holm, Jarrick, and Scott, *Humanities World Report 2015*, 13.

29. Holm, Jarrick, and Scott, *Humanities World Report 2015*, 12–13.

30. Holm, Jarrick, and Scott, *Humanities World Report 2015*, 12–13.

31. The study included the following Swedish universities: Uppsala, Lund, Gothenburg, Stockholm, Umeå, Linköping, the Royal Institute of Technology, Luleå, Karlstad, the Linnaeus University, Örebro, Mid Sweden University, and Malmö. An earlier study was repeated on February 3–4, 2024. The data presented are from the second study, which largely confirm the first study. The sample consists of descriptions of the humanities taken from the university websites (the most elaborate description that could be found, often from several pages). Verbal constructions used to describe the humanities were excerpted in their context. Translations from Swedish to English were carried out by the author.

32. Ekström and Sörlin, *Alltings mått*, 18; Östh Gustafsson, *Folkhemmets styvbarn*.

33. Ekström and Sörlin, *Alltings mått*, 19–20.

34. Woodward, "The Future of the Humanities—in the Present and in Public," 117.

35. University of Wisconsin–Madison, "About Us | UW–Madison Center for the Humanities."

36. Umeå University, "Instruktion för Wallenberg AI, Autonomous Systems and Software Program—Humanities and Society."

37. It should also be noted that in the implemented program, as of 2023, the imprint of the humanities is considerably smaller than that of social sciences and other areas,

including the number of PhD students supported. At Umeå University, for example, there are no PhD students in the Faculty of Humanities, five in the social sciences, and ten at the Department of Computing Science. Approximately 20 percent of PIs and co-PIs in the program are classified as "humanities" (based on a visual reading of a pie chart). While these numbers should be treated cautiously, they clearly point to a pattern. WASP-HS, "This Is: The Wallenberg AI, Autonomous Systems and Software Program—Humanities and Society. 2023," 15–17, 39.

38. UC Irvine, "EVOKE Lab and Studio—• Theorize • Design • Make • Provoke • Evoke • Inspire."

39. Malmö University, "Researcher Network."

40. Bardzell and Bardzell, *Humanistic HCI*, 144–145.

41. Heise, "Introduction: Planet, Species, Justice," 1.

42. Ekström and Sörlin, *Alltings mått*; Goldberg, "The War on Critical Race Theory."

43. Sörlin, "Humanities of Transformation," 288.

44. Östh Gustafsson, *Folkhemmets styvbarn*, 193–195.

45. Cuff et al., *Urban Humanities*, 15.

46. Presner, Lunenfeld, and Schnapp, "The Digital Humanities Manifesto 2.0."

47. Cuff et al., *Urban Humanities*, 8.

48. Cuff et al., *Urban Humanities*, 4.

49. Cascardi and Dear, "What Are the Urban Humanities?"

50. It is debatable whether varieties such as "public humanities" and "citizen humanities" should be included. They are not included in this analysis because they are more integrally a part of the humanities (as a method and approach) and because they do not have a strong emergent quality (see also chapter 2). Formations such as "spatial humanities" seem mostly methodological in nature.

51. Williams, "The New Humanities."

52. Svensson, "Humanities Computing as Digital Humanities"; Fiormonte, Chaudhuri, and Ricaurte, *Global Debates in the Digital Humanities*.

53. Jones et al., "The Almost Right Word."

54. Whitehead et al., *The Edinburgh Companion to the Critical Medical Humanities*; Braidotti and Oostveen, "On the Emergence and Convergence of the New Transversal Humanities," 23.

55. Cascardi and Dear, "What Are the Urban Humanities?"; Royal Holloway Centre for the GeoHumanities, "GeoHumanities Forum (About Page)."

56. Dodds and Sörlin, *Ice Humanities*.

57. Manchester University Press, "Manchester University Press—Ice Humanities (Publisher's Book Website)."

58. Bloomsbury, "The Bloomsbury Handbook to the Medical-Environmental Humanities (Information Page)."

59. Bloomsbury, "The Bloomsbury Handbook to the Medical-Environmental Humanities (Information Page)."

60. Travis et al., *Routledge Handbook of the Digital Environmental Humanities*.

61. UCLA Newsroom, "UCLA to Launch New Social Justice Curriculum with $5 Million Grant from Mellon Foundation"; UCLA Humanities Division, "UCLA Experimental Humanities Workshop on 'Humane Infrastructures'"; Bard College, "About Center for Experimental Humanities"; Open Society University Network, "Experimental Humanities Collaborative Network (EHCN) | Open Society University Network."

62. Sörlin, "Humanities of Transformation," 293.

63. Braidotti, "A Theoretical Framework for the Critical Posthumanities."

64. Braidotti, "A Theoretical Framework for the Critical Posthumanities,"43.

65. Braidotti, "A Theoretical Framework for the Critical Posthumanities," 38.

66. Braidotti, "A Theoretical Framework for the Critical Posthumanities," 53.

67. Burdick et al., *Digital Humanities*, 5.

68. Bell and Dourish, "Yesterday's Tomorrows."

69. Cascardi and Dear, "What Are the Urban Humanities?" 10.

70. Small, "Departmental Conditions and the Emergence of New Disciplines"; Balsamo, "Engineering Cultural Studies."

71. Fitzgerald and Callard, "Entangling the Medical Humanities"; Crawford and Brown, "Health Humanities."

72. Fitzgerald and Callard, "Entangling the Medical Humanities," 35.

73. Fitzgerald and Callard, "Entangling the Medical Humanities," 35.

74. Adams and Reisman, "Beyond Sparking Joy," 1404.

75. Adams and Reisman, "Beyond Sparking Joy."

76. Wenzel, "Turning over a New Leaf," 165.

77. Wenzel, "Turning over a New Leaf," 165.

78. Ratto, "Greater and Less than Human?"

79. Ferguson and Nusbaum, "Disability Studies," 78.

80. Goodley, "Dis/Entangling Critical Disability Studies," 93.

81. Corbman, "The Scholars and the Feminists."

82. Gaboury, "Critical Unmaking," 486.

83. Sundén, "On Trans-, Glitch, and Gender as Machinery of Failure"; Grundell, *Flow and Friction*.

84. Striphas, "Introduction the Long March: Cultural Studies and Its Institutionalization."

85. Halberstam, "Reflections on Queer Studies and Queer Pedagogy," 362.

86. Stone, "On Being Trans, and under the Radar."

87. Brim, *Poor Queer Studies*, 9.

88. Grossberg, *Cultural Studies in the Future Tense*, 15.

89. Hall and Jagose, "Introduction," xvi.

90. Turner, *British Cultural Studies*, 2.

91. Braidotti, "A Theoretical Framework for the Critical Posthumanities," 38.

92. Braidotti, "A Theoretical Framework for the Critical Posthumanities," 40.

93. Wolfe, *What Is Posthumanism?* 99.

94. Graef, "Muscular Humanities," 2.

95. Ginn, "Posthumanism," 414.

96. Ginn, "Posthumanism," 414.

97. Khosrowjah, "A Brief History of Area Studies and International Studies"; Duller, "Area Studies, History Of."

98. Hollinger, "Introduction," 20.

99. Smith, "The Human Touch Software of the Highest Order"; Svensson, *Big Digital Humanities*.

100. Hamraie and Fritsch, "Crip Technoscience Manifesto," 2.

101. Fritsch et al., "Introduction to Special Section on Crip Technoscience."

102. Yergeau, "Disability Hacktivism"; Mattern, "Methodolatry and the Art of Measure."

103. Yergeau, "Disability Hacktivism."

104. Hendren, *What Can a Body Do?* 17–18.

105. Hendren, *What Can a Body Do?* 9.

106. Bennett and Rosner, "The Promise of Empathy," 3–4; Irani, *Chasing Innovation*, 168–171; Ahmed, *The Cultural Politics of Emotion*, 30.

107. Hamraie, "Designing Collective Access."

108. Mills, "Body Meets World."

109. Cachia, "Introduction," 2.

110. Albrecht, Seelman, and Bury, *Handbook of Disability Studies*, 9.

111. Goodley, *Disability Studies*, 11.

112. Occidental College, "Food Studies."

113. Occidental College, "Food Studies."

114. Morris et al., "Priorities for Social Science and Humanities Research on the Challenges of Moving beyond Animal-Based Food Systems."

115. Institute for Design, Norwegian University of Science and Technology, "Design Dialogues #2."

116. The Research Council of Norway, "MEATigation"; MEATigation, "Meatigation—Mitigating Climate Change through Meat."

117. The Research Council of Norway, "MEATigation"; MEATigation, "Meatigation—Mitigating Climate Change through Meat."

118. Canali, "A Conversation with Sophia Efstathiou on the Art of Conducting Philosophy and Ethics of Scientific and Food Practices."

119. Canali, "A Conversation with Sophia Efstathiou on the Art of Conducting Philosophy and Ethics of Scientific and Food Practices."

120. The Research Council of Norway, "Evaluation of KLIMAFORSK," 13.

121. The Research Council of Norway, "Evaluation of KLIMAFORSK," 24.

122. The Research Council of Norway, "Evaluation of KLIMAFORSK," 24.

123. Stamm, "Interdisciplinarity Put to Test," 378.

124. Stamm, "Interdisciplinarity Put to Test"; Albert, Paradis, and Kuper, "Interdisciplinary Promises versus Practices in Medicine," 18; Viseu, "Integration of Social Science into Research Is Crucial."

125. Stamm, "Interdisciplinarity Put to Test," 392.

126. The Research Council of Norway, "MEATigation."

127. Svensson, "Humane Infrastructures Conversation #2."

128. The text has been edited very lightly for readability.

129. Pargman and Eriksson, "It's Not Fair!"; Eriksson and Pargman, "On the Inherent Contradictions of Teaching Sustainability at a Technical University."

130. Eriksson et al., "Addressing Students' Eco-Anxiety When Teaching Sustainability in Higher Education."

131. Pasek, Roehl, and Wellum, "Making and Meeting Online"; Pasek, "On Being Anxious about Digital Carbon Emissions."

132. Grafton and Jardine, *From Humanism to the Humanities*, 14–15.

133. Vetenskapsrådet, "Research Overview 2019," 13.

134. Vetenskapsrådet, "Research Overview 2019," 13–14.

135. Vetenskapsrådet, "Bilagor Forskningsöversikt 2019 humaniora och samhällsvetenskap," 3–6.

136. Vetenskapsrådet, "Bilagor Forskningsöversikt 2019 humaniora och samhällsvetenskap," 8–28.

137. Yale University (Office of the Provost), "Exploring the Possibilities of Artificial Intelligence at Yale."

138. The task force has two identifiable humanists that also have appointments in cognitive science and psychology.

139. Peters and Kim, "Critical Humanities Could Bring Good Trouble to AI."

140. Peters and Kim, "Critical Humanities Could Bring Good Trouble to AI."

141. Perhaps this direction is gestured at through the phrase "good trouble" in the article heading (which is, however, not used in the actual text).

142. Peters and Kim, "Critical Humanities Could Bring Good Trouble to AI."

143. During, "Friday Essay."

144. Townsend and Bradburn, "The State of the Humanities circa 2022"; Östling et al., "Kunskapens nya rörelser"; Liu et al., "What Everyone Says."

145. Pedersen, Grønvad, and Hvidtfeldt, "Methods for Mapping the Impact of Social Sciences and Humanities"; The Research Council of Norway, "Evaluation of the Humanities in Norway."

146. Sterne, "At the Interface (Course Curriculum, Spring 2023)."

147. Modern Language Association, "Guidelines for Evaluating Publicly Engaged Humanities Scholarship in Language and Literature Programs."

148. Salö, Sörlin, and Benner, "Humanvetenskapernas kunskapspolitik," 19–20; Eyerman, "Rationalizing Intellectuals," 786–787.

149. Wisselgren, "Vetenskap och /eller politik?" 112.

150. Benner, "Vetenskap och politik som gränsarbete."

151. Benner, "Vetenskap och politik som gränsarbete," 214.

152. Goldberg, "Meet Christopher Rufo"; Forsberg and Gärdebo, "Högskolan ska inte bedriva identitetspolitisk aktivism."

153. Östh Gustafsson, *Folkhemmets styvbarn*.

154. Östh Gustafsson, "Gadfly or Guide of Souls?" 134–136.

155. Östh Gustafsson, *Folkhemmets styvbarn*, 47.

156. Ekström and Sörlin, "The Integrative Humanities—and the Third Research Policy Regime," 189.

157. Östh Gustafsson, *Folkhemmets styvbarn*, 47.

158. Östh Gustafsson, *Folkhemmets styvbarn*, 425.

159. Goldberg, *The Afterlife of the Humanities*, 21–23.

160. Höög, "Idéhistoria och den klassiska bildningstraditionen. Reflektioner över humanioras framtid i det senmoderna samhället," 124.

161. Östh Gustafsson, *Folkhemmets styvbarn*, 413–415.

162. Östh Gustafsson, *Folkhemmets styvbarn*, 412.

163. Weigel, "Silicon Valley's Sixty-Year Love Affair with the Word 'Tool.'"

164. Bowker, *Science on the Run*.

165. Shapin and Schaffer, *Leviathan and the Air-Pump*, 26.

166. Svensson, "The Humanistiscope."

167. Klein, "Image of Absence"; D'Ignazio, "What Would Feminist Data Visualization Look Like?"; Drucker, *Visualization and Interpretation*; Robles-Anderson and Svensson, "One Damn Slide after Another."

168. Bowker, *Science on the Run*; Juul Søndergaard, *Staying with the Trouble through Design*; Synthesis Center, Arizona State University, "Prototyping Social Forms"; Ratto, "Greater and Less than Human?"; Grundell, *Flow and Friction*; Weizman, *Forensic Architecture*; Hamraie and Fritsch, "Crip Technoscience Manifesto."

169. Scagnetti, "A Dialogical Model for Studio Critiques in Design Education"; Arizona State University, "Humanities Lab"; Jeremijenko, "Infrastructural Recipes," (unpublished communication, October 6, 2018).

170. Spee and Jarzabkowski, "Strategy Tools as Boundary Objects," 226.

171. Star and Griesemer, "Institutional Ecology, 'Translations' and Boundary Objects."

172. I owe this example to Natalie Jeremijenko (personal communication). Cf. also Lisa Parks's notion of concealed infrastructures; Parks, "Technostruggles and the Satellite Dish," 66–67.

173. Mattern, "Unboxing the Toolkit."

174. Salö, Sörlin, and Benner, "Humanvetenskapernas kunskapspolitik," 17–18.

4: Designing Humane Infrastructures

1. Honadle, "A Capacity-Building Framework"; Eade, *Capacity-Building*.

2. Stewart, *Ordinary Affects*, 4.

3. Svensson, "Humane Infrastructures Conversation #1."

4. The text has been edited very lightly for readability.

5. Juul Søndergaard, "Troubling Design."

6. Ratto, "Critical Making."

7. Ratto and Jackson, "Reopening, Repetition and Resetting."

8. Ratto, "Greater and Less than Human?"

9. Boym, *Architecture of the Off-Modern*.

10. Sha, "The Atelier-Lab as a Transversal Machine."

11. Stone, "On Being Trans, and under the Radar"; Svensson, "Envisioning the Digital Humanities."

12. Mina and Anand, "Elephant Problems."

13. Mina and Anand, "Elephant Problems."

14. Angelo and Goh, "Out in Space."

15. Angelo and Goh, "Out in Space," 743.

16. Angelo and Goh, "Out in Space," 742.

17. Rossitto et al., "Reconsidering Scale and Scaling in CSCW Research"; Synthesis Center, Arizona State University, "Alter-Eco."

18. Ratto, "Making at the End of Nature," 28.

19. Boggs et al., "Marx, Critique, and Abolition," 514, 519.

20. Waterston, "Imagining World Solidarities for a Livable Future," 39.

21. Goldberg, *The Afterlife of the Humanities*, 25.

22. Goldberg, *The Afterlife of the Humanities*, 25.

23. I do not engage with the literature on "frame analysis" in any detail in the text, but it should be noted that that there is a variety of approaches in sociology, linguistics, media studies, and other disciplines interested in how we interpret and make sense of the world around us through "frames," including how frames are produced, employed, and negotiated.

24. Davies, "Toxic Space and Time"; Spice, "Fighting Invasive Infrastructures."

25. Hood, "In the 'Fissures of Infrastructure,'" 155.

26. Calvillo, "Atmospheric Infrastructures to Deal with the Toxic Air in a Common World."

27. Jeremijenko, "Creative Agency and the Space Race of the 21st Century."

28. Hannah and Jeremijenko, "Natalie Jeremijenko's New Experimentalism," 202.

29. Hannah and Jeremijenko, "Natalie Jeremijenko's New Experimentalism," 202.

30. Leigh, "Natalie Jeremijenko on Feral Robotic Dogs, Texting Fish and Inverted Trees."

31. Circular Economy Far North Queensland, "Museum of Natural Futures with Natalie Jeremijenko."

32. Davis, "Angela Davis: How Does Change Happen?"

33. Rentschler, "The Eavesdropper and Onlooker as Proximate Agents of Social Change," 282.

34. Parham, "Ninety-Nine Problems," 683.

35. Benjamin, *Imagination*, 19.

36. Gebru et al., "Statement from the Listed Authors of Stochastic Parrots on the 'AI Pause' Letter."

37. Lerner, *Redemptive Hope*, 26.

38. Labberton and Austin, "Righteous Protest and Public Grief, with Jeanelle Austin (Fuller Studio)."

39. Winter, "Quietly, Gradually, George Floyd Square Is Open to Traffic for the First Time since His Murder."

40. Metropolitan Council, Twin Cities metro area, "METRO D Line Construction Update- 5/6/2022."

41. Labberton and Austin, "Righteous Protest and Public Grief, with Jeanelle Austin (Fuller Studio)."

42. Patterson, "George Floyd and a Community of Care."

43. Patterson, "George Floyd and a Community of Care."

44. Chua, "Abolition Is a Constant Struggle," S-141.

45. Meet on the Streets, George Floyd Square, "What Does Justice Look Like?"

46. Arizona State University, "Introducing Jeanelle Austin, ASU's Inaugural Activist-in-Residence."

47. City of Minneapolis, "38th & Chicago Re-Envisioned."

48. City of Minneapolis, "Open House #2 Materials."

49. City of Minneapolis, "City Hosting Open House Oct. 22 for 38th and Chicago Re-Envisioned Project."

50. Berlant, *On the Inconvenience of Other People*, 25.

51. Berlant, *On the Inconvenience of Other People*, 23.

52. Berlant, "The Commons," 414.

53. There are many types of academic events. The case study centers embedded types such as conferences, symposia, and workshops, and does not engage with alternative formats such as unconferenes and hackathons. It can be noted, however, that also these alternative formats embed conditioning (e.g., see the discussion of hackathons in chapter 3 and Svensson, "Thoughts on Making Academic Events,").

54. Agre, "Hosting a Speaker."

55. Ahmed, *Living a Feminist Life*, 149.

56. Ahmed, *Living a Feminist Life*, 158.

57. Berlant, *On the Inconvenience of Other People*, 23.

58. Eriksson et al., "On the Necessity of Flying and of Not Flying."

59. Skelton, "Conferences, Conferences, Conferences?," 69–70.

60. Svensson, "From Optical Fiber to Conceptual Cyberinfrastructure."

61. McPherson, Tweet, December 6, 2012.

62. Tracy, *Colloquium*, 145.

63. Ahmed, *Living a Feminist Life*, 155.

64. Script from the "Humane Infrastructures" Conference, UCLA, March 9–10, 2020. Unpublished document.

65. I have heard several colleagues compare seminar culture across institutions where they have been active, and it is remarkable how different these can be, even when overall conditions seem similar (same country or university and the same disciplinary context).

66. Wakkary, "Nomadic Practices."

67. SCMS was founded as the Society of Cinematologists in 1959. The first annual meeting took place at the New York University Faculty Club in April 1960; Society for Cinema and Media Studies, "SCMS Organizational History."

68. Society for Cinema and Media Studies, "SCMS Annual Conference."

69. Mitchell, "A Few Thoughts on Improving the SCMS Conference."

70. Society for Cinema and Media Studies, "SCMS 2019 Conference Program," 18.

71. Society for Cinema and Media Studies, "SCMS 2019 Conference Program," 18.

72. Society for Cinema and Media Studies, "SCMS 2020 Cancellation Official Statement."

73. Society for Cinema and Media Studies, "Transcript: SCMS 2022 Conference Decision (February 3)."; Society for Cinema and Media Studies, "2022 SCMS Conference Update (Video)."

74. Society for Cinema and Media Studies, "2022 SCMS Conference Update (Video)."

75. Society for Cinema and Media Studies. "SCMS Members Meeting 2022, 2022 Virtual Conference (Presentation)."

76. Svensson, "Virtual Worlds as Arenas for Language Learning"; Svensson, "The Digital Humanities as a Humanities Project"; Svensson, *Big Digital Humanities*.

77. Wershler-Henry, Emerson, and Parikka, *The Lab Book*; Riley, "Lab/Studio"; MacCannell, "Institutes and Institutions."

78. Sha, "The Atelier-Lab as a Transversal Machine," 64.

79. Balsamo, *Designing Culture*, 162–163.

80. Irani, "Design Thinking."

81. Collins, Evans, and Gorman, "Trading Zones and Interactional Expertise," 9–12.

82. Galison, *Image and Logic*.

83. Star and Griesemer, "Institutional Ecology, 'Translations' and Boundary Objects," 393.

84. Stamm, "Interdisciplinarity Put to Test," 385–386; Lyall et al., "Key Success Factors in the Quest for Interdisciplinary Knowledge," 2.

85. Stirling, "Direction, Distribution and Diversity!"

86. Arnold, "From Caring to Creating"; Cook, "Immateriality and Its Discontents."

87. Svensson, *Big Digital Humanities*, 184–194.

88. Graham and Cook, *Rethinking Curating*, 10; Big Digital Humanities, 190.

89. Mina and Anand, "Elephant Problems."

90. Sörlin, "Så blev Stockholm en världsstad i kampen mot miljöförstörelsen."

91. Fins, "Understanding and Utilizing the Convening Power of Ethics Consultation"; Bedore, "The Convening Power of Food as Growth Machine Politics."

92. Sörlin, "Så blev Stockholm en världsstad i kampen mot miljöförstörelsen"; Sörlin and Paglia, "Stockholm and 1972—Capital of Environmental Memory."

93. Pratt, "Arts of the Contact Zone," 34.

94. Lymer, *The Work of Critique in Architectural Education.*

95. Barrett, "Studio Critiques of Student Art," 30–32; Utaberta et al., "Upgrading Education Architecture by Redefining Critique Session in Design Studio," 46.

96. Interlocutors (apart from Sha Xin Wei) include Sally Kitch, Adam Nocek, Lisa Yin Han, Gaymon Bennett, all Arizona State University, as well as Mia Phillipson, Uppsala University, Michael Goodhart, University of Pittsburgh, and Fanny Wonu Veys, the National Museum of World Cultures in the Netherlands.

97. Sha and Svensson, "May 5 ASU-SCAS Workshop Building Humanistic and Humane Capacity."

98. I was prompted to consider the critique of "clinic" as a model more profoundly by Sally Kitch and Lisa Yin Han, both Arizona State University.

99. The idea to conceive of the IC as a service came from Sha Xin Wei, building on a joint conversation about the liminal qualities necessary for an operation of this kind.

100. This point was made strongly by Gaymon Bennett, Arizona State University, in a conversation.

Conclusion: Toward Humane Infrastructures

1. Regeringskansliet (Swedish Government), "Nya styrelser för 30 universitet och högskolor"; Reslow et al., "SD: Stoppa Genusindoktrinering"; Andersson, "Stoppa finansieringen av genusteori inom högre utbildning"; Governor of Florida, "Governor Ron DeSantis Signs Legislation to Protect Floridians from Discrimination and Woke Indoctrination."

2. Cornford, "Inverting Resolution."

3. Cornford, "Inverting Resolution," 108.

4. Cullberg, "Så designas de hemlösa bort från stadsmiljöerna"; Benjamin, *Imagination*, chap. 3; Samorani et al., "Overbooked and Overlooked."

5. Svensson, "Humanistisk infrastruktur (main author: Svensson)."

6. Letter from Svensson to the Swedish Research Council ("Def."), April 17, 2018. Unpublished document.

7. Vetenskapsrådet, "Forskning & innovation för framtiden."

Bibliography

Aagaard Jensen, Sanne. "Connecting the Alliance: Communications Infrastructure on the NATO Agenda in the 1950s." In *Historicizing Infrastructure*, edited by Andreas Marklund and Mogens Rüdiger, 183–210. Aalborg: Aalborg University Press, 2017.

Adams, Zoe, and Anna Reisman. "Beyond Sparking Joy: A Call for a Critical Medical Humanities." *Academic Medicine* 94, no. 10 (October 2019): 1404.

Agre, Phil. "Hosting a Speaker: A Guide for Graduate Students." UCLA, 2001. https://pages.gseis.ucla.edu/faculty/agre/hosting.html.

Ågren, Maria, and Elisabet Engdahl. "Gör konsekvensanalys innan tidskriftsstödet avskaffas." *Curie*, November 27, 2013.

Ahmed, Sara. *Living a Feminist Life*. Durham, NC: Duke University Press, 2017.

Ahmed, Sara. *The Cultural Politics of Emotion*. 2nd ed. Edinburgh: Edinburgh University Press, 2014.

Albert, Mathieu, Elise Paradis, and Ayelet Kuper. "Interdisciplinary Promises versus Practices in Medicine: The Decoupled Experiences of Social Sciences and Humanities Scholars." *Social Science and Medicine* 126 (February 2015): 17–25.

Albrecht, Gary L., Katherine Delores Seelman, and Michael Bury. *Handbook of Disability Studies*. Thousand Oaks: SAGE Publications, 2001.

Alemu, Kassaye Sintayehu. "The Meaning, Idea and History of University/Higher Education in Africa: A brief Literature Review." *FIRE: Forum for International Research in Education* 4, no. 3 (2018): 210–227.

All European Academies (ALLEA). "Going Digital: Creating Change in the Humanities (ALLEA E-Humanities Working Group Report)." Berlin, 2015. https://www.allea.org/wp-content/uploads/2015/07/Going-Digital_digital-version.pdf.

Altshuler, Alan. "Review: Infrastructure Investment." *Journal of Policy Analysis and Management* 8, no. 3 (1989): 505–508.

American Council of Learned Societies (ACLS). "Our Cultural Commonwealth: The Report of the American Council of Learned Societies Commission on Cyberinfrastructure for the Humanities and Social Sciences." 2006. https://www.acls.org/wp-content/uploads/2021/11/Our-Cultural-Commonwealth.pdf.

American Society of Civil Engineers. "2017 Infrastructure Report Card." Accessed June 13, 2018. http://www.infrastructurereportcard.org.

Amoore, Louise. "Machine Learning Political Orders." *Review of International Studies* 49, no. 1 (2023): 20–36.

Amréus, Lars, Ann Follin, Karin Grönvall, Joakim Malmström, and Karin Åström Iko. "Digitalisera kulturarvet för demokratins skull" (DN Debatt). *Dagens Nyheter*, January 3, 2021.

Anand, Nikhil. *Hydraulic City: Water and the Infrastructures of Citizenship in Mumbai*. Durham, NC: Duke University Press, 2017.

Andersson, Tobias. "Stoppa finansieringen av genusteori inom högre utbildning (Tobias Andersson, Sweden Democrats)." Bill to Parliament (Motion 2019/20:813), 2019.

Angelo, Hillary, and Kian Goh. "Out in Space: Difference and Abstraction in Planetary Urbanization." *International Journal of Urban and Regional Research (IJURR)* 45, no. 4 (2021): 732–744.

Aradau, Claudia. "Security That Matters: Critical Infrastructure and Objects of Protection." *Security Dialogue* 41, no. 5 (October 2010): 491–514.

Arbona, Javier, Roderic Crooks, Kathleen Frederickson, Sage Gerson, Anirban Gupta-Nigam, Adriana Johnson, Kameron Sanzo, Ameeth Vijay, Janet Walker, Salvador Zárate, and Susan Zieger. "Reading for Infrastructure." Fabrica, UCHRI Foundry, February 2021. https://uchri.org/foundry/reading-for-infrastructure/.

Arizona State University. "Humanities Lab." Accessed August 22, 2022. https://humanities.lab.asu.edu/.

Arizona State University. "Introducing Jeanelle Austin, ASU's Inaugural Activist-in-Residence." *ASU News*, January 4, 2022. https://news.asu.edu/20220104-introducing-jeanealle-austin-asus-inaugural-activist-residence.

Arizona State University. "Like a Medical Center for the Planet." *ASU News*, January 13, 2022. https://news.asu.edu/20220113-solutions-medical-center-planet.

Arizona State University. "PlanetWorks—Designing and Managing the Future of Planet Earth." PlanetWorks. Accessed May 29, 2021. https://sustainability-innovation.asu.edu/planetworks/.

Arnold, Ken. "From Caring to Creating: Curators Change Their Spots." In *The International Handbooks of Museum Studies: Volume 2*, edited by Conal McCarthy, 317–339. Chichester, UK: John Wiley, 2015.

Aselmeyer, Norman. "Capturing the Zeitgeist of Infrastructure. Reflections from the Transregional Academy on 'Infrastructures, Regions, and Urbanization,' 29 September to 7 October 2018, Singapore." TRAFO—Blog for Transregional Research (blog). January 31, 2019. https://trafo.hypotheses.org/17723.

Association of Universities and Colleges of Canada, Canadian Association of University Teachers, and National Consortium of Scientific and Educational Societies. "Putting

Knowledge to Work: Sustaining Canada as an Innovative Society. An Action Agenda. (Brief Presented to Federal Government)." October 18, 1996.

Atkins, Daniel E., Kelvin K. Droegemeier, Stuart I. Feldman, Hector García-Molina, Michael L. Klein, David G. Messerschmitt, Paul Messina, Jeremiah P. Ostriker, and Margaret H. Wright. "Revolutionizing Science and Engineering through Cyberinfrastructure: Report of the National Science Foundation Blue-Ribbon Advisory Panel on Cyberinfrastructure." Arlington, VA: National Science Foundation. 2003. https://citeseerx.ist.psu.edu/document?doi=841a9bb02faa96203b5058252640b260a5538e3e.

Atkinson, M., R. Baxter, P. Brezany, O. Corcho, M. Galea, M. Parsons, D. Snelling, and J. van Hemert. "Data-Driven Research in the Humanities—the DARIAH Research Infrastructure." In *The Data Bonanza: Improving Knowledge Discovery in Science, Engineering, and Business*, edited by Malcolm Atkinson, Rob Baxter, Peter Brezany, Oscar Corcho, Michelle Galea, Mark Parsons, David Snelling, and Jano van Hemert, 417–430. New York: IEEE, 2013.

Australian Government. "2016 National Research Infrastructure Roadmap." 2017. https://www.education.gov.au/download/3658/2016-national-research-infrastructure-roadmap/5326/document/pdf.

Ayers, Edward L. "Where the Humanities Live." *Daedalus* 138, no. 1 (Winter 2009): 24–34.

Baca, Murtha. "Getty Voices: Rethinking Art History." Getty Iris (blog). March 4, 2013. http://blogs.getty.edu/iris/getty-voices-rethinking-art-history/.

Balsamo, Anne. *Designing Culture: The Technological Imagination at Work*. Durham, NC: Duke University Press, 2011.

Balsamo, Anne. "Engineering Cultural Studies: The Postdisciplinary Adventures of Mindplayers, Fools, and Others." In *Doing Science + Culture*, edited by Roddey Reid and Sharon Traweek, 259–274. New York: Routledge, 2000.

Banfill, Jonathan, Todd Presner, and Maite Zubiaurre. "Urban Humanities Pedagogy: The Classroom, Education, and New Humanities." *Boom* 6, no. 3 (September 1, 2016): 120–128.

Barclay, Glen St. John, and Caroline Turner. *Humanities Research Centre: A History of the First 30 Years of the HRC at The Australian National University*. Canberra: ANU E Press, 2004.

Bard College. "About Center for Experimental Humanities." Experimental Humanities. Accessed May 13, 2021. https://eh.bard.edu/about-eh/.

Bardzell, Jeffrey, and Shaowen Bardzell. *Humanistic HCI*. Synthesis Lectures on Human-Centered Informatics. San Rafael: Morgan & Claypool, 2015.

Barrett, Terry. "Studio Critiques of Student Art: As They Are, as They Could Be with Mentoring." *Theory into Practice* 39, no. 1 (2000): 29–35.

Bear, Laura. "'Alternatives' to Austerity: A Critique of Financialized Infrastructure in India and Beyond." *Anthropology Today* 33, no. 5 (2017): 3–7.

Bear, Laura. *Navigating Austerity: Currents of Debt along a South Asian River*. Stanford, CA: Stanford University Press, 2015.

Becker, Sophia, Dirk von Schneidemesser, Alexandre Caseiro, Katharina Götting, Seán Schmitz, and Erika von Schneidemesser. "Pop-up Cycling Infrastructure as a Niche Innovation for Sustainable Transportation in European Cities: An Inter- and Transdisciplinary Case Study of Berlin." *Sustainable Cities and Society* 87 (December 2022): 104168.

Bedore, Melanie. "The Convening Power of Food as Growth Machine Politics: A Study of Food Policymaking and Partnership Formation in Baltimore." *Urban Studies* 51, no. 14 (2014): 2979–2995.

Bell, Genevieve, and Paul Dourish. "Yesterday's Tomorrows: Notes on Ubiquitous Computing's Dominant Vision." *Personal and Ubiquitous Computing* 11, no. 2 (January 2007): 133–143.

Bengtsson, Stefan, Pam Fredman, Karin Dahlman-Wright, Sigbritt Karlsson, Helen Dannetun, Torbjörn von Schantz, Astrid Söderbergh Widding, Peter Högberg, and Eva Åkesson. "Forskningen hotas." *Upsala Nya Tidning*, January 7, 2017.

Benjamin, Ruha. *Imagination: A Manifesto*. New York: W. W. Norton, 2024.

Benner, Mats. "Big Science in a Small Country: Constraints and Possibilities of Research Policy." In *In Pursuit of a Promise: Perspectives on the Political Process to Establish the European Spallation Source (ESS) in Lund, Sweden*, edited by Olof Hallonsten, 159–172. Lund: Arkiv förlag, 2012.

Benner, Mats. "Vetenskap och politik som gränsarbete." In *Humanvetenskapernas verkningar*, edited by Linus Salö, 200–220. Stockholm: Dialogos Förlag, 2021.

Bennett, Cynthia L., and Daniela K. Rosner. "The Promise of Empathy: Design, Disability, and Knowing the 'Other.'" In *Proceedings of the 2019 CHI Conference on Human Factors in Computing Systems*, edited by Stephen Brewster and Geraldine Fitzpatrick, 1–13. CHI '19. New York: Association for Computing Machinery, 2019.

Berend, I. T. "Development of the Infrastructure—in Historical Perspective." *Acta Oeconomica* 17, no. 3–4 (1976): 245–256.

Berg, Johanna, Rickard Domeij, Jens Edlund, Gunnar Eriksson, Per Fallgren, David House, Eva Lindström, Erik Magnusson Petzell, Zofia Malisz, Susanne Nylund Skog, and Jenny Öqvist. "Making Archival Speech Recordings Accessible for Research: A Report from the Tilltal Project." *Svenska landsmål och svenskt folkliv* 141 (2019): 171–178.

Berlant, Lauren. *On the Inconvenience of Other People*. Durham, NC: Duke University Press Books, 2022.

Berlant, Lauren. "The Commons: Infrastructures for Troubling Times." *Environment and Planning D: Society and Space* 34, no. 3 (2016): 393–419.

Besedovsky, Natalia, Fritz-Julius Grafe, Hanna Hilbrandt, and Hannes Langguth. "Time as Infrastructure." *City* 23, no. 4–5 (2019): 580–588.

Bhattacharyya, Sayan. "Epistemically Produced Invisibility." In *Global Debates in the Digital Humanities*, edited by Domenico Fiormonte, Sukanta Chaudhuri, and Paola Ricaurte, 3–14. Minneapolis: University of Minnesota Press, 2022.

Bloomsbury. "The Bloomsbury Handbook to the Medical-Environmental Humanities (Information Page)." Accessed July 24, 2022. https://www.bloomsbury.com/us/bloomsbury-handbook-to-the-medicalenvironmental-humanities-9781350197305/.

Bod, Rens. *A New History of the Humanities: The Search for Principles and Patterns from Antiquity to the Present*. Oxford: Oxford University Press, 2013.

Boggs, Abigail, Eli Meyerhoff, Nick Mitchell, and Zach Schwartz-Weinstein. "Marx, Critique, and Abolition: Higher Education as Infrastructure." In *The Palgrave International Handbook of Marxism and Education*, edited by Richard Hall, Inny Accioly, and Krystian Szadkowski, 509–535. Cham: Springer, 2023.

Bolliger, Isabel K., and Alexandra Griffiths. "The Introduction of ESFRI and the Rise of National Research Infrastructure Roadmaps in Europe." In *Big Science and Research Infrastructures in Europe*, edited by Katharina C. Cramer and Olof Hallonsten, 101–127. Cheltenham, UK: Edward Elgar, 2020.

Borgman, Christine L. *Scholarship in the Digital Age: Information, Infrastructure, and the Internet*. Cambridge, MA: MIT Press, 2007.

Bowker, Geoffrey C. "Data Flakes: An Afterword to *'Raw Data' Is an Oxymoron*." In *"Raw Data" Is an Oxymoron*, edited by Lisa Gitelman, 167–171. Cambridge, MA: MIT Press, 2013.

Bowker, Geoffrey C. *Science on the Run. Information Management and Industrial Geophysics at Schlumberger, 1920–1940*. Cambridge, MA: MIT Press, 1994.

Bowker, Geoffrey C., Karen Baker, Florence Millerand, and David Ribes. "Toward Information Infrastructure Studies: Ways of Knowing in a Networked Environment." In *International Handbook of Internet Research*, edited by Jeremy Huisinger, Lisbeth Klastrup, and Matthew Allen, 97–117. Dordrecht: Springer, 2010.

Bowker, Geoffrey C., and Susan Leigh Star. *Sorting Things Out: Classification and Its Consequences*. Cambridge, MA: MIT Press, 1999.

Boyle, Philip J., and Shannon T. Speed. "From Protection to Coordinated Preparedness: A Genealogy of Critical Infrastructure in Canada." *Security Dialogue* 49, no. 3 (2018): 217–231.

Boym, Svetlana. *Architecture of the Off-Modern*. New York: Princeton Architectural Press, 2008.

Braidotti, Rosi. "A Theoretical Framework for the Critical Posthumanities." *Theory, Culture and Society* 36, no. 6 (2019): 31–61.

Braidotti, Rosi, and Daan F. Oostveen. "On the Emergence and Convergence of the New Transversal Humanities." In *The Edinburgh Companion to the New European Humanities*, edited by Rosi Braidotti, Hiltraud Casper-Hehne, Marjan Ivković, and Daan F. Oostveen, 23–46. Edinburgh: Edinburgh University Press, 2024.

Bratton, Benjamin. "Planetary Sapience." *Noema*, June 17, 2021. https://www.noemamag.com/planetary-sapience.

Bratton, Benjamin. "The Terraforming." *KALEIDOSCOPE*, no. 37 (2020–2021).

Brim, Matt. *Poor Queer Studies: Confronting Elitism in the University*. Durham, NC: Duke University Press, 2020.

Brontë, Lydia, ed. "A Guide to Humanities Centers in the United States." Working Papers, Rockefeller Foundation, New York, 1980.

Brown, Deidre, and George Nicholas. "Protecting Indigenous Cultural Property in the Age of Digital Democracy: Institutional and Communal Responses to Canadian First Nations and Māori Heritage Concerns." *Journal of Material Culture* 17, no. 3 (2012): 307–324.

Brown, Susan, Tanya E. Clement, Laura Mandell, Deb Verhoeven, and Jacque Wernimont. "Creating Feminist Infrastructure in the Digital Humanities." In *Digital Humanities 2016: Conference Abstracts. Jagiellonian University and Pedagogical University, Kraków*, edited by Maciej Eder and Jan Rybicki, 47–50. Kraków: ADHO, 2016.

Bullard, Robert D. "Anatomy of Environmental Racism and the Environmental Justice Movement." In *Confronting Environmental Racism: Voices from the Grassroots*, edited by Robert D. Bullard, 15–39. Boston: South End Press, 1993.

Burchardt, Marian, and Stefan Höhne. "The Infrastructures of Diversity: Materiality and Culture in Urban Space—An Introduction." *New Diversities* 17, no. 2 (2015): 1–13.

Burdick, Anne, Johanna Drucker, Peter Lunenfeld, Todd Presner, and Jeffrey Schnapp. *Digital_Humanities*. Cambridge, MA: MIT Press, 2012.

Busch, Lawrence. *Standards: Recipes for Reality*. Cambridge, MA: MIT Press, 2011.

Cachia, Amanda. "Introduction: Committed to Change—Ten Years of Creative Access." In *Curating Access*, edited by Amanda Cachia, 1–14. New York: Routledge, 2023.

Calvillo, Nerea. "Atmospheric Infrastructures to Deal with the Toxic Air in a Common World." *Ardeth*, 5 (2019): 184–197.

Canada Foundation for Innovation. "Our Mandate." Accessed December 23, 2023. https://www.innovation.ca/about/mandate.

Canali, Stefano. "A Conversation with Sophia Efstathiou on the Art of Conducting Philosophy and Ethics of Scientific and Food Practices." *SPSP Newsletter #20 (Society for Philosophy of Science in Practice)*, December 2023.

Carnochan, W. Bliss. “Institutes and the Conscience of the King.” In *The Institutional Impact of Institutes, University of California Humanities Research Institute, April 29.30, 1988—Papers for Discussion*, edited by Ralph Cohen, 1–5. Irvine: University of California Humanities Research Institute, 1988.

Carroll, Lucy, Lena Nguyen, and EJ Panaligan. “UCLA Fails to Account for Student Homelessness despite Needing Data to Take Action.” *Daily Bruin*, January 13, 2020. https://dailybruin.com/2020/01/13/ucla-fails-to-account-for-student-homelessness-despite-needing-data-to-take-action/.

Carse, Ashley. “Keyword: Infrastructure: How a Humble French Engineering Term Shaped the Modern World.” In *Infrastructures and Social Complexity: A Companion*, edited by Penny Harvey, Casper Bruun Jensen, and Atsuro Morita, 27–39. New York: Routledge, 2017.

Cascardi, Anthony, and Michael Dear. “What Are the Urban Humanities?” *Boom* 6, no. 3 (2016): 4–11.

Caswell, Michelle, and Samip Mallick. “Against Precarity: Towards a Community-Based Notion of Fiscal Sustainability.” *Medium*, July 30, 2018. https://medium.com/community-archives/against-precarity-towards-a-community-based-notion-of-fiscal-sustainability-815d1d889309.

Chalmers University of Technology. “A Big Investment to Make Chalmers Equal.” July 6, 2018. https://www.chalmers.se/en/news/Pages/big-investment-to-make-Chalmers-equal.aspx.

Chalmers University of Technology. “GENIE Gender Report 2018.” March 6, 2020. https://www.chalmers.se/en/about-chalmers/organisation-and-governance/equality/genie-gender-initiative-for-excellence/.

Chalmers University of Technology. “GENIE Gender Report 2019.” August 31, 2020. https://www.chalmers.se/en/about-chalmers/organisation-and-governance/equality/genie-gender-initiative-for-excellence/.

Chalmers University of Technology. “Research Infrastructure at Chalmers.” Accessed November 13, 2020. https://www.chalmers.se/en/researchinfrastructure/Pages/default.aspx.

Chalmers University of Technology and KTH Royal Institute of Technology. “Climate Framework for Higher Education Institutions—Guidelines.” 2019. https://www.kth.se/polopoly_fs/1.929789.1600783144!/191001_guidelines_Klimatramverk_ENG_anp.pdf.

Chandler, James. “Vulnerability Without and Within.” *Profession*, November 2014. https://profession.mla.org/vulnerability-without-and-within/.

Chiland, Elijah. “City Officials Give up on Playa Del Rey Road Diets.” *Curbed LA*, October 18, 2017.

Choate, Pat, and Susan Walter. *America in Ruins: Beyond the Public Works Pork Barrel.* Washington, DC: Council of State Planning Agencies, 1981.

Choate, Pat, and Susan Walter. *America in Ruins: The Decaying Infrastructure*. Durham, NC: Duke Press Paperbacks, 1983.

Chua, Charmaine. "Abolition Is a Constant Struggle: Five Lessons from Minneapolis." *Theory and Event* 23, no. 4 (2020): S127–S147.

Circular Economy Far North Queensland. "Museum of Natural Futures with Natalie Jeremijenko," n.d. Accessed January 3, 2023. https://www.circulareconomyfnq.org.au/museum-of-natural-futures-by-natalie-jeremijenko.

City of Minneapolis. "38th & Chicago Re-Envisioned." Accessed January 10, 2023. https://www.minneapolismn.gov/government/projects/38th-chicago-ave/.

City of Minneapolis. "City Hosting Open House Oct. 22 for 38th & Chicago Re-Envisioned Project." October 12, 2022. https://web.archive.org/web/20221210082450/https://www.minneapolismn.gov/news/2022/october/38th-and-chicago-re-envisioned-/.

City of Minneapolis. "Open House #2 Materials." October 22, 2022. https://www.minneapolismn.gov/media/-www-content-assets/documents/Open-House-2---38th-Chicago-Re-Envisioned,-October-22,-2022.pdf.

CLARIN. "CLARIN ERIC." Accessed January 26, 2024. https://www.clarin.eu/.

Clark, William. "On the Dialectical Origins of the Research Seminar." *History of Science* 27, no. 2 (1989): 111–154.

Collier, Stephen J., and Andrew Lakoff. *The Government of Emergency: Vital Systems, Expertise, and the Politics of Security*. Princeton, NJ: Princeton University Press, 2021.

Collier, Stephen J., and Andrew Lakoff. "The Vulnerability of Vital Systems: How 'Critical Infrastructure' Became a Security Problem." In *Securing "the Homeland": Critical Infrastructure, Risk and (in)Security*, edited by Myriam Dunn Cavelty and Kristian Søby Kristensen, 17–39. London: Routledge, 2008.

Collier, Stephen J., and Andrew Lakoff. "Vital Systems Security: Reflexive Biopolitics and the Government of Emergency." *Theory, Culture and Society* 32, no. 2 (March 1, 2015): 19–51.

Collins, Harry, Robert Evans, and Michael E. Gorman. "Trading Zones and Interactional Expertise." In *Trading Zones and Interactional Expertise: Creating New Kinds of Collaboration*, edited by Michael E. Gorman, 7–23. Cambridge, MA: MIT Press, 2010.

Collinson, Stephen. "The Fight to Define Infrastructure Could Change America." CNN. Accessed April 6, 2021. https://edition.cnn.com/2021/04/06/politics/joe-biden-infrastructure-us-debate/index.html.

Congressional Budget Office, Congress of the United States. "Public Works Infrastructure: Policy Considerations for the 1980s." CBO Study, April 1983. https://www.cbo.gov/sites/default/files/98th-congress-1983-1984/reports/doc20-entire.pdf.

Cook, Sarah. "Immateriality and Its Discontents: An Overview of Main Models and Issues for Curating New Media." In *New Media in the White Cube and Beyond: Curatorial*

Models for Digital Art, edited by Christiane Paul, 26–49. Berkeley: University of California Press, 2008.

Corbman, Rachel. "The Scholars and the Feminists: The Barnard Sex Conference and the History of the Institutionalization of Feminism." *Feminist Formations* 27, no. 3 (2015): 49–80.

Cornell University. "About Us | Society for the Humanities." Accessed August 10, 2024. https://societyhumanities.as.cornell.edu/about-us.

Cornford, Stephen. "Inverting Resolution: Accounting for the Planetary Cost of Earth Observation." *Journal of Visual Culture* 22, no. 1 (March 21, 2023): 93–110.

Costanza-Chock, Sasha. *Design Justice: Community-Led Practices to Build the Worlds We Need*. Cambridge, MA: MIT Press, 2020.

Coulson, Jonathan, Paul Roberts, and Isabelle Taylor. *University Planning and Architecture: The Search for Perfection*. 2nd ed. New York: Routledge, 2015.

Cramer, Katharina C., Olof Hallonsten, Isabel K. Bolliger, and Alexandra Griffiths. "Big Science and Research Infrastructures in Europe: History and Current Trends." In *Big Science and Research Infrastructures in Europe*, edited by Katharina C. Cramer and Olof Hallonsten, 1–26. Cheltenham, UK: Edward Elgar, 2020.

Crawford, P., and Brian J. Brown. "Health Humanities: A Democratizing Future beyond Medical Humanities." In *Routledge Handbook of the Medical Humanities*, edited by Alan Bleakley, 401–409. New York: Routledge, 2020.

Crosby, Andrew. "The Racialized Logics of Settler Colonial Policing: Indigenous 'Communities of Concern' and Critical Infrastructure in Canada." *Settler Colonial Studies* 11, no. 4 (2021): 411–430.

Cuff, Dana, Anastasia Loukaitou-Sideris, Todd Presner, Maite Zubiaurre, and Jonathan Jae-An Crisman. *Urban Humanities: New Practices for Reimagining the City*. Cambridge, MA: MIT Press, 2020.

Cullberg, Jonas. "Så designas de hemlösa bort från stadsmiljöerna." *ETC*, November 10, 2014.

Cybersecurity and Infrastructure Security Agency CISA (America's Cyber Defense Agency). "Critical Infrastructure Security and Resilience." Accessed June 9, 2023. https://www.cisa.gov/topics/critical-infrastructure-security-and-resilience.

Dahlström, Mats, Joacim Hansson, and Ulrika Kjellman. "'As We May Digitize'—Institutions and Documents Reconfigured." *LIBER Quarterly* 21, no. 3–4 (2012): 455–474.

Dalingwater, Louise. "Post-New Public Management (NPM) and the Reconfiguration of Health Services in England." *Observatoire de La Société Britannique*, no. 16 (2014): 51–64.

DARIAH-EU. "DARIAH | Digital Research Infrastructure for the Arts and Humanities." Accessed January 26, 2024. https://www.dariah.eu/.

Davidson, Cathy. "The Futures of Scholarly Publishing." *Journal of Scholarly Publishing* 35, no. 3 (2004): 129–142.

Davies, Mark. "Corpus of Historical American English (COHA)." 2010. https://www.english-corpora.org/coha/.

Davies, Thom. "Toxic Space and Time: Slow Violence, Necropolitics, and Petrochemical Pollution." *Annals of the American Association of Geographers* 108, no. 6 (2018): 1537–1553.

Davis, Angela. "Angela Davis: How Does Change Happen?" UC Davis Lecture (video recording), 2006. https://www.youtube.com/watch?v=Pc6RHtEbiOA.

Deb, Sopan. "Trump Proposes Eliminating the Arts and Humanities Endowments." *The New York Times*, March 15, 2017.

Dempsey, Paul Stephen. "The Rise and Fall of the Interstate Commerce Commission: The Tortuous Path from Regulation to Deregulation of America's Infrastructure." *Marquette Law Review* 95, no. 4 (2012): 1151–1190.

D'Ignazio, Catherine. "What Would Feminist Data Visualization Look Like?" MIT Center for Civic Media, 2015. https://civic.mit.edu/2015/12/01/feminist-data-visualization/.

D'Ignazio, Catherine, and Lauren F. Klein. *Data Feminism*. Cambridge, MA: MIT Press, 2020.

Directorate-General for Research and Innovation (European Commission). "Research Infrastructures Make Science Happen (Infographic)." Publications Office of the European Union, 2019.

DiSalvo, Carl. *Adversarial Design*. Cambridge, MA: MIT Press, 2012.

Dodds, Klaus, and Sverker Sörlin, eds. *Ice Humanities: Living, Working, and Thinking in a Melting World*. Manchester: Manchester University Press, 2022.

Donald J. Trump website. "Infrastructure (Donald J. Trump's Vision)." 2016. https://web.archive.org/web/20161129080933/https://www.donaldjtrump.com/policies/an-americas-infrastructure-first-plan/.

Donaldson, Ian. "The Humanities Research Centre, The Australian National University." In *The Institutional Impact of Institutes, University of California Humanities Research Institute, April 29.30, 1988—Papers for Discussion*, edited by Ralph Cohen, 1–6. Irvine: University of California Humanities Research Institute, 1988.

Drake, Jarrett. *#ArchivesSoWhite in the Words of Jarrett Drake. Issues and Advocacy*. 2016. https://issuesandadvocacy.wordpress.com/2016/04/19/archivessowhite-in-the-words-of-jarrett-drake/.

Drucker, Johanna. "Blind Spot: Information Visualization and Art History." In *The Routledge Companion to Digital Humanities and Art History*, edited by Kathryn Brown, 18–31. New York: Routledge, 2020.

Drucker, Johanna. *Visualization and Interpretation: Humanistic Approaches to Display*. Cambridge, MA: MIT Press, 2020.

Drucker, Johanna, and Patrik Svensson. "The Why and How of Middleware." *Digital Humanities Quarterly* 10, no. 2 (2016).

Duller, Matthias. "Area Studies, History Of." In *International Encyclopedia of the Social and Behavioral Sciences*, edited by James D. Wright, 949–954. Oxford: Elsevier, 2015.

Dumit, Joseph. "Writing the Implosion: Teaching the World One Thing at a Time." *Cultural Anthropology* 29, no. 2 (2014): 344–362.

Dunn Cavelty, Myriam. "Like a Phoenix from the Ashes: The Reinvention of Critical Infrastructure Protection as Distributed Security." In *Securing "the Homeland": Critical Infrastructure, Risk and (In)Security*, edited by Myriam Dunn Cavelty and Kristian Søby Kristensen, 40–62. New York: Routledge, 2008.

During, Simon. "Friday Essay: Simon During on the Demoralisation of the Humanities, and What Can Be Done about It." *The Conversation*, July 14, 2022.

During, Simon. "The Idea of the Humanities." Unpublished paper (speech), 2017. https://www.academia.edu/34926361/The_idea_of_the_humanities_2017_.

Eade, Deborah. *Capacity-Building: An Approach to People-Centred Development*. Oxford: Oxfam, 1997.

Eby, John. "Humanities Hall Ready to Welcome Students." *Clemson News* (Clemson University), August 15, 2022. https://news.clemson.edu/humanities-hall-ready-to-welcome-students/.

Edwards, Paul N. "Infrastructure and Modernity: Scales of Force, Time, and Social Organization in the History of Sociotechnical Systems." In *Modernity and Technology*, edited by Thomas J. Misa, Philip Brey, and Andrew Feenberg, 185–225. Cambridge, MA: MIT Press, 2003.

Edwards, Paul N. *A Vast Machine: Computer Models, Climate Data, and the Politics of Global Warming*. Cambridge, MA: MIT Press, 2010.

Edwards, Paul N., Steven J. Jackson, Geoffrey C. Bowker, and Cory P Knobel. "Understanding Infrastructure: Dynamics, Tensions, and Design. Report of a Workshop on 'History and Theory of Infrastructure, Lessons for New Scientific Cyberinfrastructures.'" Ann Arbor: Deep Blue, 2007. https://deepblue.lib.umich.edu/handle/2027.42/49353.

Efstathiou, Sophia. "Is It Possible to Give Scientific Solutions to Grand Challenges? On the Idea of Grand Challenges for Life Science Research." *Studies in History and Philosophy of Biological and Biomedical Sciences* 56 (April 2016): 48–61.

Ehn, Billy, and Orvar Löfgren. *Hur blir man klok på universitetet?* Lund: Studentlitteratur, 2004.

Ekström, Anders. "Forging the Integrative Humanities: Policies and Prospects." In *The Humanities and the Modern Politics of Knowledge. The Impact and Organization of the Humanities in Sweden, 1850–2020*, edited by Anders Ekström and Hampus Östh Gustafsson, 275–289. Amsterdam: Amsterdam University Press, 2022.

Ekström, Anders, and Sverker Sörlin. *Alltings mått: humanistisk kunskap i framtidens samhälle*. Stockholm: Norstedts, 2012.

Ekström, Anders, and Sverker Sörlin. "The Integrative Humanities—and the Third Research Policy Regime." In *Smart Policies for Societies in Transition: The Innovation Challenge of Inclusion, Resilience and Sustainability*, edited by Mats Benner, Göran Marklund, and Sylvia Schwaag Serger, 189–212. Cheltenham, UK: Edward Elgar, 2022.

Ellison, Julie. "Guest Column—The New Public Humanists." *PMLA* 2, no. 128 (2013): 289–298.

Eriksson, Elina, Anne-Kathrin Peters, Daniel Pargman, Björn Hedin, Minna Laurell-Thorslund, and Sandra Sjöö. "Addressing Students' Eco-Anxiety When Teaching Sustainability in Higher Education." In *Proceedings: 2022 International Conference on ICT for Sustainability (ICT4S)*, Plovdiv, Bulgaria, 88–98. New York: IEEE, 2022.

Eriksson, Elina, and Daniel Pargman. "On the Inherent Contradictions of Teaching Sustainability at a Technical University." In *Digital Technology and Sustainability: Engaging the Paradox*, edited by Mike Hazas and Lisa Nathan, 154–165. New York: Routledge, 2018.

Eriksson, Elina, Daniel Pargman, Markus Robèrt, and Jarmo Laaksolahti. "On the Necessity of Flying and of Not Flying: Exploring How Computer Scientists Reason about Academic Travel." In *Proceedings of the 7th International Conference on ICT for Sustainability (ICT4S)*, edited by Ruzanna Chitchyan and Daniel Schien, 18–26. New York: Association for Computing Machinery, 2020.

Eriksson, Pär, and Svante Barck-Holst. "Politik för skydd av kritisk infrastruktur i EU och Sverige—en jämförande analys." Totalförsvarets forskningsinstitut (FOI), FOI-R—1793—SE, 2005. https://www.foi.se/rest-api/report/FOI-R--1793--SE.

ESFRI Working Group on Monitoring of Research Infrastructures Performance. "Monitoring of Research Infrastructures Performance." 2019. https://www.esfri.eu/sites/default/files/ESFRI_WG_Monitoring_Report.pdf.

Estrada, Gilbert. "If You Build It, They Will Move: The Los Angeles Freeway System and the Displacement of Mexican East Los Angeles, 1944–1972." *Southern California Quarterly* 87, no. 3 (October 1, 2005): 287–315.

European Commission. "About Research Infrastructures | Infrastructures—Research and Innovation." Accessed July 24, 2018. https://ec.europa.eu/research/infrastructures/index.cfm?pg=about.

European Commission. "Communication from the Commission of 12 December 2006 on a European Programme for Critical Infrastructure Protection [COM(2006) 786 Final—Official Journal C 126 of 7.6.2007]." 2006. https://eur-lex.europa.eu/EN/legal-content/summary/european-programme-for-critical-infrastructure-protection.html.

European Commission. "European Open Science Cloud." October 28, 2016. https://ec.europa.eu/digital-single-market/en/european-open-science-cloud.

European Commission. "Horizon Europe." December 20, 2023. https://research-and-innovation.ec.europa.eu/funding/funding-opportunities/funding-programmes-and-open-calls/horizon-europe_en.

European Commission. "Horizon Europe Work Programme 2023–2024, 3. Research Infrastructures (European Commission Decision C (2024) 2371 of 17 April 2023)." April 17, 2024. https://ec.europa.eu/info/funding-tenders/opportunities/docs/2021-2027/horizon/wp-call/2023-2024/wp-3-research-infrastructures_horizon-2023-2024_en.pdf.

European Commission. "Statutes of the European Spallation Source ERIC." 2015. https://europeanspallationsource.se/sites/default/files/downloads/2017/09/ERIC Statutes.pdf.

European Commission. "Towards a European Research Area," Communication from the Commission to The Council, the European Parliament, the Economic and Social Committee and the Committee of the Regions January 18, 2000. https://eur-lex.europa.eu/LexUriServ/LexUriServ.do?uri=COM:2000:0006:FIN:EN:PDF.

European Commission, and Canada Foundation for Innovation. "International Conference on Research Infrastructures Proceedings Report June 1–3, 2021." Canada, June 1, 2021. http://icri2021.ca/wp-content/uploads/2021/09/ICRI-2021-Proceedings-Report.pdf.

European Commission. Directorate-General for Research and Innovation. *Future of Scholarly Publishing and Scholarly Communication: Report of the Expert Group to the European Commission.* Luxembourg: Publications Office of the European Union, 2019.

European Commission, Directorate-General for Research and Innovation. "Supporting the Transformative Impact of Research Infrastructures on European Research—Report of the High-Level Expert Group to Assess the Progress of ESFRI and Other World Class Research Infrastructures towards Implementation and Long-Term Sustainability." June 24, 2020. https://research-and-innovation.ec.europa.eu/knowledge-publications-tools-and-data/publications/all-publications/supporting-transformative-impact-research-infrastructures-european-research_en.

European Open Science Cloud (EOSC). "About Page, EOSC." Accessed February 25, 2021. https://eosc-portal.eu/about/eosc.

European Open Science Cloud (EOSC). "EOSCpilot Demonstrator: TEXTCROWD." Accessed February 26, 2021. https://eosc-portal.eu/eoscpilot-demonstrator-textcrowd.

European Open Science Cloud (EOSC). "TEXTCROWD (EOSCPilot)." Accessed February 26, 2021. https://eoscpilot.eu/science-demos/textcrowd.

European Parliament and the European Council. Decision No 182/1999/EC of the European Parliament and of the Council of 22 December 1998 concerning the fifth framework programme of the European Community for research, technological development and demonstration activities (1998 to 2002), *Official Journal of the European Union*, L 26, 42 (February 1, 1999). https://eur-lex.europa.eu/legal-content/EN/TXT/PDF/?uri=OJ:L:1999:026:FULL.

European Science Foundation. "Research Infrastructures in the Digital Humanities," ESF Science Policy Briefing 42. September 2011. http://archives.esf.org/fileadmin/Public_documents/Publications/spb42_RI_DigitalHumanities.pdf.

European Science Foundation, and European Cooperation in Science and Technology. "Responses to Environmental and Societal Challenges for Our Unstable Earth (RESCUE)." ESF Forward Look—ESF-COST "Frontier of Science" Joint Initiative. Strasbourg and Brussels, 2011. http://archives.esf.org/fileadmin/Public_documents/Publications/rescue.pdf.

European Spallation Source. "European Spallation Source." Accessed February 13, 2021. https://europeanspallationsource.se/about.

European Strategy Forum on Research Infrastructures. "ESFRI Roadmap 2021, Projects and Landmarks." 2021. https://roadmap2021.esfri.eu/projects-and-landmarks/.

European Strategy Forum on Research Infrastructures. "European Roadmap for Research Infrastructures. Report 2006," 2006. https://www.esfri.eu/sites/default/files/esfri_roadmap_2006_en.pdf.

European Strategy Forum on Research Infrastructures. "Roadmap 2018: Strategy Report on Research Infrastructures." 2018. http://roadmap2018.esfri.eu/media/1066/esfri-roadmap-2018.pdf.

European Strategy Forum on Research Infrastructures. "Roadmap for European Research Infrastructures: Report of the Social Sciences and Humanities Roadmap Working Group." 2006. https://www.esfri.eu/sites/default/files/ssh-rwg-roadmap-report-2006_en_0.pdf.

European Strategy Forum on Research Infrastructures. "Social Sciences and Humanities: Roadmap Working Group Report 2008." 2008. https://www.esfri.eu/sites/default/files/ssh_report_2008_en_0.pdf.

European Strategy Forum on Research Infrastructures. "Strategy Report on Research Infrastructures: Roadmap 2016." European Strategy Forum on Research Infrastructures, 2016. https://www.esfri.eu/sites/default/files/20160308_ROADMAP_single_page_LIGHT.pdf.

European Strategy Forum on Research Infrastructures, Long-Term Sustainability Working Group. "Long-Term Sustainability of Research Infrastructures." 2017. https://ec.europa.eu/research/infrastructures/pdf/esfri/publications/esfri_scripta_vol2.pdf.

Eyerman, Ron. "Rationalizing Intellectuals: Sweden in the 1930s and 1940s." *Theory and Society* 14, no. 6 (1985): 777–807.

Ferguson, Philip M., and Emily Nusbaum. "Disability Studies: What Is It and What Difference Does It Make?" *Research and Practice for Persons with Severe Disabilities* 37, no. 2 (2012): 70–80.

Fidler, Bradley, and Andrew L. Russell. "Financial and Administrative Infrastructure for the Early Internet: Network Maintenance at the Defense Information Systems Agency." *Technology and Culture* 59, no. 4 (2018): 899–924.

Fins, Joseph J. "Understanding and Utilizing the Convening Power of Ethics Consultation." *AMA Journal of Ethics* 18, no. 5 (2016): 540–545.

Fiormonte, Domenico, Sukanta Chaudhuri, and Paola Ricaurte, eds. *Global Debates in the Digital Humanities*. Minneapolis: University of Minnesota Press, 2022.

Fitzgerald, Des, and Felicity Callard. "Entangling the Medical Humanities." In *The Edinburgh Companion to the Critical Medical Humanities*, edited by Anne Whitehead, Angela Woods, Sarah Atkinson, Jane Macnaughton, and Jennifer Richards, 35–49. Edinburgh: Edinburgh University Press, 2016.

Fitzpatrick, Kathleen. *Planned Obsolescence: Publishing, Technology, and the Future of the Academy*. New York: NYU Press, 2011.

Fleischer, Matthew. "Bulldoze L.A. freeways." *Los Angeles Times*, June 24, 2020. https://www.latimes.com/opinion/story/2020-06-24/bulldoze-la-freeways-racism-monument.

Forsberg Lundell, Fanny, and Johan Gärdebo. "Högskolan ska inte bedriva identitetspolitisk aktivism." *Göteborgs-Posten*, July 3, 2022.

Fritsch, Kelly, Aimi Hamraie, Mara Mills, and David Serlin. "Introduction to Special Section on Crip Technoscience." *Catalyst: Feminism, Theory, Technoscience* 5, no. 1 (April 1, 2019): 1–10.

Frykholm, Joel. "Critical Thinking and the Humanities: A Case Study of Conceptualizations and Teaching Practices at the Section for Cinema Studies at Stockholm University." *Arts and Humanities in Higher Education* 20, no. 3 (2021): 253–273.

Fulmer, Jeffrey E. "What in the World is Infrastructure?" *Infrastructure Investor*, July/August 2009, 30–32.

Gaboury, Jacob. "Critical Unmaking: Toward a Queer Computation." In *The Routledge Companion to Media Studies and Digital Humanities*, edited by Jentery Sayers, 483–491. New York: Routledge, 2018.

Gabrys, Jennifer. *Digital Rubbish: A Natural History of Electronics*. Ann Arbor, MI: University of Michigan Press, 2011.

Gabrys, Jennifer. "Programming Nature as Infrastructure in the Smart Forest City." *Journal of Urban Technology* 29, no. 1 (2022): 13–19.

Galison, Peter. *Image and Logic: A Material Culture of Microphysics*. Chicago: University of Chicago Press, 1997.

Gärdebo, Johan, Mattias Höjer, David Nilsson, and Kristoffer Soldal. "Klimatforskarna förvärrar utsläpp." *Svenska Dagbladet*, December 4, 2015, sec. Debatt. https://www.svd.se/klimatforskarna-forvarrar-utslapp.

GBIF. "Global Biodiversity Information Facility. Free and Open Access to Biodiversity Data." Accessed June 26, 2018. https://www.gbif.org/.

Gebru, Timnit, Emily M. Bender, Angelina McMillan-Major, and Margaret Mitchell. "Statement from the Listed Authors of Stochastic Parrots on the 'AI Pause' Letter." The DAIR Institute, March 31, 2023. https://www.dair-institute.org/blog/letter-statement-March2023.

Giamatti, A. Bartlett. "Introductory Remarks." In *On the Occasion of the Inauguration of the Whitney Humanities Center*. New Haven, CT: Yale University, 1981.

Gilliland, Anne J., and Michelle Caswell. "Records and Their Imaginaries: Imagining the Impossible, Making Possible the Imagined." *Archival Science* 16, no. 1 (2016): 53–75.

Ginn, Franklin. "Posthumanism." In *The Edinburgh Companion to Animal Studies*, edited by Lynn Turner, Undine Sellbach, and Ron Broglio, 413–429. Edinburgh: Edinburgh University Press, 2018.

Goh, Kian. "Planning the Green New Deal: Climate Justice and the Politics of Sites and Scales." *Journal of the American Planning Association* 86, no. 2 (2020): 188–195.

Goldberg, David Theo. *Dread: Facing Futureless Futures*. Cambridge, UK: Polity, 2021.

Goldberg, David Theo. "Meet Christopher Rufo—Leader of the Incoherent Right-Wing Attack on 'Critical Race Theory.'" *Salon*, August 1, 2021.

Goldberg, David Theo. *The Afterlife of the Humanities*. Irvine: University of California Humanities Research Institute, 2014. https://issuu.com/uchri/docs/afterlife/.

Goldberg, David Theo. "The War on Critical Race Theory." *Boston Review*, May 2, 2021.

Goodheart, Eugene. "Reflections on the Institutional Impact of Institutes." In *The Institutional Impact of Institutes, University of California Humanities Research Institute, April 29.30, 1988—Papers for Discussion*, edited by Ralph Cohen, 1–4. Irvine: University of California Humanities Research Institute, 1988.

Goodley, Dan. *Disability Studies: An Interdisciplinary Introduction*. Thousand Oaks, CA: SAGE Publications, 2017.

Goodley, Dan. "Dis/Entangling Critical Disability Studies." In *Culture—Theory—Disability*, edited by Anne Waldschmidt, Hanjo Berressem, and Moritz Ingwersen, 81–109. Bielefeld: transcript Verlag, 2017.

Gordon, Constance, and Kyle Byron. "Sweeping the City: Infrastructure, Informality, and the Politics of Maintenance." *Cultural Studies* 35, no. 4–5 (2021): 854–875.

Government Bill (House of Commons) C-93 (35–2)—Royal Assent—An Act to implement certain provisions of the budget tabled in Parliament on February 18, 1997—Parliament of Canada, Pub. L. No. C–93 (1997). https://www.parl.ca/documentviewer/en/35-2/bill/C-93/royal-assent/page-29#3.

Government of Canada, Legislative Services. "Budget Implementation Act, 1997." Accessed December 11, 2021. https://laws.justice.gc.ca/eng/acts/B-9.81/page-1.html.

Governor of Florida, Ron DeSantis. "Governor Ron DeSantis Signs Legislation to Protect Floridians from Discrimination and Woke Indoctrination (News Releases)." April 22, 2022. https://www.flgov.com/2022/04/22/governor-ron-desantis-signs-legislation-to-protect-floridians-from-discrimination-and-woke-indoctrination/.

Graef, Ortwin de. "Muscular Humanities." Panel presentation presented at the ECHIC Conference, Humanities Now: Global Encounters, University of Macerata, April 21–22, 2016.

Grafton, Anthony, and Lisa Jardine. *From Humanism to the Humanities: Education and the Liberal Arts in Fifteenth and Sixteenth-Century Europe*. Cambridge, MA: Harvard University Press, 1986.

Graham, Beryl, and Sarah Cook. *Rethinking Curating: Art After New Media*. Cambridge, MA: MIT Press, 2010.

Granberg, Anders. "The ESS Project as a Generator of Conflict and Collaboration: An Assessment of the Official Picture of Costs and Benefits and the Research-Community Response." In *In Pursuit of a Promise: Perspectives on the Political Process to Establish the European Spallation Source (ESS) in Lund, Sweden*, edited by Olof Hallonsten, 109–157. Lund: Arkiv, 2012.

GRIDH, University of Gothenburg, Chalmers University of Technology, and Riksdagens öppna data, Swedish Government. "Hitta och läs riksdagsmotioner." https://riksdagsmotioner.dh.gu.se/.

Grossberg, Lawrence. *Cultural Studies in the Future Tense*. Durham, NC: Duke University Press, 2010.

Groves, Christopher, Max Munday, and Natalia Yakovleva. "Fighting the Pipe: Neoliberal Governance and Barriers to Effective Community Participation in Energy Infrastructure Planning." *Environment and Planning C: Government and Policy* 31, no. 2 (2013): 340–356.

Grundell, Vendela. *Flow and Friction: On the Tactical Potential of Interfacing with Glitch Art*. Stockholm: Art and Theory, 2016.

Guajardo Soto, Guillermo. "What Is Infrastructure? Origins, Turns and Continuities of the Concept." *ARQ (Santiago)*, no. 114 (2023): 4–15.

Guédon, Cécile. "A Curious Halo—Digital Salon Podcast." Digital Salon, August 24, 2020. https://digitalsalonpodcast.org/A-Curious-Halo.

Halberstam, Judith. "Reflections on Queer Studies and Queer Pedagogy." *Journal of Homosexuality* 45, no. 2–4 (2003): 361–364.

Hall, Donald E., and Annamarie Jagose. "Introduction." In *The Routledge Queer Studies Reader*, edited by Donald E. Hall and Annamarie Jagose, xiv–xx. London: Routledge, 2013.

Hallinan, Blake, and James N. Gilmore. "Infrastructural Politics amidst the Coils of Control." *Cultural Studies* 35, no. 4–5 (2021): 617–640.

Hallonsten, Olof. "Introduction. In Pursuit of a Promise." In *In Pursuit of a Promise: Perspectives on the Political Process to Establish the European Spallation Source (ESS) in Lund, Sweden*, edited by Olof Hallonsten, 11–19. Lund: Arkiv, 2012.

Hallonsten, Olof. "Research Infrastructures in Europe: The Hype and the Field." *European Review* 28, no. 4 (2020): 617–635.

Hammar, Isak, and Johan Östling. "Introduction: The Circulation of Knowledge and the History of Humanities." *History of Humanities* 6, no. 2 (2021): 595–602.

Hamraie, Aimi. "Designing Collective Access: A Feminist Disability Theory of Universal Design." *Disability Studies Quarterly* 33, no. 4 (2013).

Hamraie, Aimi, and Kelly Fritsch. "Crip Technoscience Manifesto." *Catalyst: Feminism, Theory, Technoscience* 5, no. 1 (2019): 1–33.

Han, Lisa Y. "Sonic Pipelines at the Seafloor." *Media + Environment* 3, no. 2 (2021).

Hanke, Steve H. "Shaky Foundation, Overbuilt Solution." Reason.com, April 1983. https://reason.com/archives/1983/04/01/shaky-foundation-overbuilt-sol.

Hannah, Dehlia, and Natalie Jeremijenko. "Natalie Jeremijenko's New Experimentalism." In *Anthropocene Feminism*, edited by Richard Grusin, 197–219. Minneapolis: University of Minnesota Press, 2017.

Hansson, Göran K. "Yttrande från Kungl. Vetenskapsakademien inför den kommande forskningspolitiska propositionen 2016." Kungl. Vetenskapsakademien, October 21, 2015. Accessed June 15, 2018. https://www.kva.se/app/uploads/2015/10/globalassets-vetenskapsamhallet-remisser-2015-yttrandeforskningspropositionen2016.pdf.

Hansson, Sven Ove. "Konsten att vara vetenskaplig." KTH Royal Institute of Technology, 2007. https://people.kth.se/~soh/konstenatt.pdf.

Harvard School of Engineering and Applied Sciences. "Model Shows Solar Geoengineering May Be Surprisingly Effective in Alleviating Impacts of Global Warming on Crops." Accessed May 31, 2021. https://www.seas.harvard.edu/news/2021/05/model-shows-solar-geoengineering-may-be-surprisingly-effective-alleviating-impacts.

Hasenöhrl, Ute, and Jan-Henrik Meyer. "The Energy Challenge in Historical Perspective." *Technology and Culture* 61, no. 1 (2020): 295–306.

HASS Committee on the Humanities Library, MIT. "Final Report." Humanities Library Report, Cambridge, MA, August 2001.

Head, Alison. "Envisioning Biblotheca 2.0: One of the Most Exciting Design Tasks of Our Era." Project Information Literacy (Interview with Jeffrey Schnapp), 2012. https://projectinfolit.org/smart-talk-interviews/envisioning-biblotheca-2-0-one-of-the-most-exciting-design-tasks-of-our-era/.

Head, Brian W. *Wicked Problems in Public Policy: Understanding and Responding to Complex Challenges*. Cham: Springer, 2022.

Heise, Ursula K. "Introduction: Planet, Species, Justice—and the Stories We Tell about Them." In *The Routledge Companion to the Environmental Humanities*, edited by Ursula K. Heise, Jon Christensen, and Michelle Niemann, 1–10. New York: Routledge, 2017.

Heise, Ursula K. "Terraforming for Urbanists." *Novel* 49, no. 1 (May 2016): 10–25.

Henderson, Theo. "We the Unhoused." Accessed August 12, 2020. https://soundcloud.com/user-369990655.

Hendren, Sara. *What Can a Body Do?: How We Meet the Built World*. New York: Riverhead Books, 2020.

Henke, Christopher R., and Benjamin Sims. *Repairing Infrastructures: The Maintenance of Materiality and Power*. Cambridge, MA: MIT Press, 2020.

Hill, Dan. "Another Green World." *Medium*, April 7, 2020. https://medium.com/slowdown-papers/10-another-green-world-f70e7bcf904c.

Hollinger, David A. "Introduction." In *The Humanities and the Dynamics of Inclusion since World War II*, edited by David A. Hollinger, 1st ed., 1–22. Baltimore: Johns Hopkins University Press, 2006.

Holm, Poul, Arne Jarrick, and Dominic Scott. *Humanities World Report 2015*. Basingstoke, UK: Palgrave Macmillan, 2015.

Honadle, Beth Walter. "A Capacity-Building Framework: A Search for Concept and Purpose." *Public Administration Review* 41, no. 5 (1981): 575–580.

Hood, Kate Lewis. "In the 'Fissures of Infrastructure': Poetry and Toxicity in 'Garbage Arcadia.'" *Environmental Humanities* 13, no. 1 (2021): 136–158.

Höög, Victoria. "Idéhistoria och den klassiska bildningstraditionen. Reflektioner över humanioras framtid i det senmoderna samhället." In *Hvor kommer idéhistorien fra? Tematiske og teoretiske brytninger i idéhistoriefaget*, edited by Sven Arntzen and Otto M. Christensen, 121–136. Kristiansand: Höjskoleförlaget, 2003.

Horton, Zachary. *The Cosmic Zoom: Scale Knowledge, and Mediation*. Chicago: University of Chicago Press, 2021.

Huang, Josie. "Protesters Killed a Plan for a Koreatown Homeless Shelter. But That's Just the Beginning of the Story." *LAist*, August 1, 2019. https://laist.com/projects/2019/koreatown-shelter-protest/.

Hughes, Thomas Parke. *Networks of Power: Electrification in Western Society, 1880–1930*. Baltimore: Johns Hopkins University Press, 1983.

Huynh, Kayla. "UW–Madison Building Projects: 5 to Watch." *The Cap Times*, August 6, 2023.

Infrastructure and Projects Authority, UK. "National Infrastructure Delivery Plan 2016 to 2021." 2016. https://assets.publishing.service.gov.uk/media/5a7f2f1540f0b6230268df36/2904569_nidp_deliveryplan.pdf.

Institute for Advanced Study Princeton. "IAS Welcomes Over 250 Scholars for 2022–23 Academic Year—Press Release | Institute for Advanced Study." September 14, 2022. https://www.ias.edu/news/2022-23-welcome.

Institutt for design, Norwegian University of Science and Technology. "Design Dialogues #2." Facebook, June 10, 2021.

Irani, Lilly. *Chasing Innovation: Making Entrepreneurial Citizens in Modern India*. Princeton, NJ: Princeton University Press, 2019.

Irani, Lilly. "'Design Thinking': Defending Silicon Valley at the Apex of Global Labor Hierarchies." *Catalyst: Feminism, Theory, Technoscience* 4, no. 1 (2018): 1–19.

Ivković, Marjan, and Đurđa Trajković. "Public Humanities Today: Between Community Engagement and Social Critique." In *The Edinburgh Companion to the New European Humanities*, edited by Rosi Braidotti, Hiltraud Casper-Hehne, Marjan Ivković, and Daan F. Oostveen, 106–121. Edinburgh: Edinburgh University Press, 2024.

Jackson, Steven J. "Rethinking Repair." In *Media Technologies: Essays on Communication, Materiality, and Society*, edited by Tarleton Gillespie, Pablo J. Boczkowski, and Kirsten A. Foot, 221–239. Cambridge, MA: MIT Press, 2014.

Jeremijenko, Natalie. "Creative Agency and the Space Race of the 21st Century; Towards a Museum of Natural Futures." In *Proceedings of the 2016 ACM Conference on Designing Interactive Systems—DIS '16*, edited by Marcus Foth, 3–4. New York: Association for Computing Machinery, 2016.

Johnson, Adriana, and Susan Zieger. "Reading for Infrastructure, Infrastructure for Reading (UCHRI Short-Term Collaborative Residency)." Accessed February 15, 2021. https://uchri.org/awards/reading-for-infrastructure-infrastructure-for-reading/.

Jones, Therese, Michael Blackie, Rebecca Garden, and Delese Wear. "The Almost Right Word: The Move from Medical to Health Humanities." *Academic Medicine* 92, no. 7 (2017): 932–935.

Juckes, Martin, Michael Bane, Jennifer Bulpett, Katie Cartmell, Miranda MacFarlane, Molly MacRae, Alex Owen, Charlotte Pascoe, and Poppy Townsend. "Sustainability in Digital Research Infrastructure: UKRI Net Zero DRI Scoping Project Final Technical Report." 2023. https://zenodo.org/records/8199984.

Juul Søndergaard, Marie Louise. *Staying with the Trouble through Design: Critical-Feminist Design of Intimate Technology*. Aarhus: Aarhus University, 2018.

Juul Søndergaard, Marie Louise. "Troubling Design: A Design Program for Designing with Women's Health." *ACM Transactions on Computer–Human Interaction (TOCHI)* 27, no. 4 (2020): 1–36.

Kahn, Paul W. "On Humanities and Human Rights." In *The Humanities and Public Life*, edited by Peter Brooks and Hilary Jewett, 116–122. New York: Fordham University Press, 2014.

Kaldewey, David. "The Grand Challenges Discourse: Transforming Identity Work in Science and Science Policy." *Minerva* 56, no. 2 (2018): 161–182.

Kaplan, E. Ann. "What Impact Should, or Can a Humanities Institute Have on the Organization of Knowledge in the University?" In *The Institutional Impact of Institutes, University of California Humanities Research Institute, April 29.30, 1988—Papers for Discussion*, edited by Ralph Cohen, 1–9. Irvine: University of California Humanities Research Institute, 1988.

Karasti, Helena, and Karen S. Baker. "Infrastructuring for the Long-Term: Ecological Information Management." In *Proceedings of the 37th Annual Hawaii International Conference on System Sciences*, 2004. https://www.computer.org/csdl/proceedings-article/hicss/2004/01265077/12OmNwwuE1v.

Karlsohn, Thomas. "The Academic Seminar as Emotional Community." *Nordic Journal of Studies in Educational Policy* 2016, no. 2–3 (2016).

Karpf, Juanita. "The Last Dramatic Instructional Innovation? The Chalkboard and Music Education in a Historical Prospective." *Journal of Historical Research in Music Education* 34, no. 1 (2012): 62–81.

Kaufmann-Buhler, Jennifer. *Open Plan: A Design History of the American Office*. London: Bloomsbury, 2021.

Kennedy, John F. "John F. Kennedy: Remarks of Welcome to President Ayub Khan of Pakistan at Quonset Point, Rhode Island." American Presidency Project, UC Santa Barbara, 1962.

Khosrowjah, Hossein. "A Brief History of Area Studies and International Studies." *Arab Studies Quarterly* 33, no. 3/4 (2011): 131–142.

Kitchin, Rob. *The Data Revolution: Big Data, Open Data, Data Infrastructures and Their Consequences*. London: SAGE Publications, 2014.

Klein, Lauren F. "Image of Absence: Archival Silence, Data Visualization, and James Hemings." *American Literature* 85, no. 4 (2013): 661–688.

Knödlseder, Jürgen, Sylvie Brau-Nogué, Mickael Coriat, Philippe Garnier, Annie Hughes, Pierrick Martin, and Luigi Tibaldo. "Estimate of the Carbon Footprint of Astronomical Research Infrastructures." *Nature Astronomy* 6 (2022): 503–513.

Koh, Adeline, and Roopika Risam. "Open Thread: The Digital Humanities as a Historical 'Refuge' from Race/Class/Gender/Sexuality/Disability?" *Postcolonial Digital Humanities* (blog). 2013.

KTH Royal Institute of Technology. "Klimatramverket." Accessed August 6, 2024. https://www.kth.se/om/miljo-hallbar-utveckling/klimatramverket-1.903489.

Kungl. Vetenskapsakademien (Royal Swedish Academy of Sciences). "Forskningspropositionen 2008/09:50 rörande MAX IV och ESS (letter to the Government)." November 24, 2008. https://essochmaxiv.wordpress.com/wp-content/uploads/2009/04/kva-brev.pdf.

Kungliga biblioteket. "Årsredovisning 2019." 2020. https://www.kb.se/samverkan-och-utveckling/kbs-uppdrag/arsredovisningar.html.

Kungliga biblioteket. "Årsredovisning 2020." 2021. https://www.kb.se/samverkan-och-utveckling/kbs-uppdrag/arsredovisningar.html.

Kungliga biblioteket. "Årsredovisning 2021." 2022. https://www.kb.se/samverkan-och-utveckling/kbs-uppdrag/arsredovisningar.html.

Kungliga biblioteket. "Årsredovisning 2022." 2023. https://www.kb.se/samverkan-och-utveckling/kbs-uppdrag/arsredovisningar.html.

Kungliga biblioteket. "Färdplan 2023–2025." 2023. https://www.kb.se/download/18.2958abae1862d609fbd3ca2/1678438019602/F%C3%A4rdplan-2023-2025.pdf.

Kungliga biblioteket. "Svenska Dagstidningar (Swedish Newspapers)." Accessed June 1, 2023. https://tidningar.kb.se/.

LA Times Editorial Board. “Editorial: NIMBYs Stop a Hotel from Housing Sick Homeless People in Orange County.” *Los Angeles Times*, April 10, 2020.

Laak, Dirk van. *Lifelines of Our Society: A Global History of Infrastructure*. Translated by Erik Butler. Cambridge, MA: MIT Press, 2023.

Labberton, Mark, and Jeanelle Austin. “Righteous Protest and Public Grief, with Jeanelle Austin (Fuller Studio).” April 13, 2021. https://fullerstudio.fuller.edu/podcast/righteous-protest-and-public-grief-with-jeanelle-austin/.

Larkin, Brian. “The Politics and Poetics of Infrastructure.” *Annual Review of Anthropology* 42 (October, 2013): 327–343.

Lawhon, Mary, Alexander Follmann, Boris Braun, Natasha Cornea, Clemens Greiner, Prince Guma, Timos Karpouzoglou, Javier Revilla Diez, Seth Schindler, Sophie Schramm, Franziska Sielker, Gideon Tups, Sumit Vij, and Peter Dannenbergl. “Making Heterogeneous Infrastructure Futures in and beyond the Global South.” *Futures* 154 (December, 2023): 1–15.

Lee, Charlotte P., Paul Dourish, and Gloria Mark. “The Human Infrastructure of Cyberinfrastructure.” In *Proceedings of the 2006 20th Anniversary Conference on Computer Supported Cooperative Work*, edited by Pamela Hinds and David Martin, 483–492. CSCW ’06. New York: Association for Computing Machinery, 2006.

Lee, Sheng. “New State-of-the-Art Humanities Building to Replace Susan B. Davis Residence Hall, Former Zoe Bayliss Co-Op.” *The Badger Herald*, October 27, 2023.

Leef, Georges. “The Case for Abolishing the National Endowment for the Humanities Just Got Stronger (Opinion Piece).” *Forbes*, July 3, 2017.

Leigh, Andrew. “Natalie Jeremijenko on Feral Robotic Dogs, Texting Fish and Inverted Trees (The Good Life Podcast with Andrew Leigh).” Interview by Andrew Leigh, May 24, 2018. https://www.andrewleigh.com/natalie_jeremijenko_tgl.

Lerner, Akiba J. *Redemptive Hope: From the Age of Enlightenment to the Age of Obama*. New York City: Fordham University Press, 2015.

Liddle, Hannah. “The Rise of Public Scholarship.” *University Affairs*, November 9, 2023. https://universityaffairs.ca/features/feature-article/the-rise-of-public-scholarship/.

Liu, Alan. “Critical Infrastructure Studies (MLA Panel Description, Liu as Presider).” Modern Language Association Conference 2018, New York City, 2018. https://criticalinfrastructure.hcommons.org/session-description/.

Liu, Alan, Abigail Droge, Scott Kleinman, Lindsay Thomas, Dan C. Baciu, and Jeremy Douglass. “What Everyone Says: Public Perceptions of the Humanities in the Media.” *Dædalus* 151, no. 3 (2022): 19–39.

Livingstone, David N. *Putting Science in Its Place: Geographies of Scientific Knowledge*. Chicago: University of Chicago Press, 2003.

Lopreite, Débora, and Joan Murphy. “The Canada Foundation for Innovation as Patron and Regulator.” In *Research and Innovation Policy: Changing Federal Government-University*

Relations, edited by G. Bruce Doern and Christopher Stoney, 123–147. Toronto: University of Toronto Press, 2009.

Los Angeles Department of Transportation. "LADOT Annual Report, January 1, 2022–December 31, 2022." City of Los Angeles, 2023. https://ladot.lacity.gov/sites/default/files/documents/annual-report-2022_2023.pdf.

Loukissas, Yanni Alexander. *All Data Are Local: Thinking Critically in a Data-Driven Society*. Cambridge, MA: MIT Press, 2019.

Lundquist, Magnus. "Här stängslar Gävle Energi bort värmen för en hemlös." *Gefle Dagblad*, February 21, 2020.

Lyall, Catherine, Ann Bruce, Wendy Marsden, and Laura Meagher. "Key Success Factors in the Quest for Interdisciplinary Knowledge." The ESRC Centre for Social and Economic Research on Innovation in Genomics (Innogen), March 2011. https://www.pure.ed.ac.uk/ws/portalfiles/portal/12623512/Key_success_factors_in_the_quest_for_interdisciplinary_knowledge.pdf.

Lymer, Gustav. *The Work of Critique in Architectural Education*. Gothenburg: Acta Universitatis Gothoburgensis, 2010.

MacCannell, Juliet Flower. "Institutes and Institutions: The Critical Difference." In *The Institutional Impact of Institutes, University of California Humanities Research Institute, April 29.30, 1988—Papers for Discussion*, edited by Ralph Cohen, 1–5. Irvine: University of California Humanities Research Institute, 1988.

Malmö University. "Researcher Network: Collaborative Future Making." May 3, 2018. https://web.archive.org/web/20180503082617/http://medea.mah.se/alternative-future-making/.

Manchester University Press. "Manchester University Press—Ice Humanities (Publisher's Book Website)." 2022. https://manchesteruniversitypress.co.uk/9781526157775/ice-humanities/.

Marx, Leo. *The Machine in the Garden: Technology and the Pastoral Ideal in America*. Anniversary edition. Oxford: Oxford University Press, 1964.

Masco, Joseph. "The Age of Fallout." *History of the Present* 5, no. 2 (2015): 137–168.

Matres, Inés, Mila Oiva, and Mikko Tolonen. "In Between Research Cultures—The State of Digital Humanities in Finland." *Informaatiotutkimus* 37, no. 2 (2018).

Mattern, Shannon. "Interfacing Urban Intelligence." *Places Journal*, April 2014. https://placesjournal.org/article/interfacing-urban-intelligence/.

Mattern, Shannon. "Library as Infrastructure." *Places Journal*, June 2014. https://placesjournal.org/article/library-as-infrastructure/.

Mattern, Shannon. "Methodolatry and the Art of Measure." *Places Journal*, November 2013. https://placesjournal.org/article/methodolatry-and-the-art-of-measure/.

Mattern, Shannon. "Unboxing the Toolkit." *Toolshed*, July 9, 2021. https://tool-shed.org/unboxing-the-toolkit/.

McCabe, Bret. "Hopkins Humanities Center Celebrates 50 Years as Home to a Diverse Intellectual Community." *The Hub (John Hopkins University)*, March 22, 2016. https://hub.jhu.edu/2016/03/22/hopkins-humanities-fiftieth-conference/.

McLean, Ellen. "Jane Ohlmeyer Is an Academic Unafraid of Trinity's Political Tightrope." *The University Times*, February 23, 2018. http://www.universitytimes.ie/2018/02/jane-ohlmeyer-is-an-academic-unafraid-of-trinitys-political-tightrope/.

McPherson, Tara. "Designing for Difference." *Differences* 25, no. 1 (2014): 177–188.

McPherson, Tara. *Feminist in a Software Lab: Difference + Design*. Cambridge, MA: Harvard University Press, 2018.

McPherson, Tara. "Theory/Practice: Lessons Learned from Feminist Film Studies." In *The Routledge Companion to Media Studies and Digital Humanities*, edited by Jentery Sayers, 9–17. New York: Routledge, 2018.

McPherson, Tara [@tmcphers]. "#mediaplaces2012 I take it as a v. good sign that over the last 2 days I find myself wishing that the 'jr.' scholars were doing the keynotes." Twitter, December 6, 2012.

MEATigation. "Meatigation—Mitigating Climate Change through Meat." Accessed July 20, 2022. https://meatigation.no/.

Meet on the Streets, George Floyd Square. "What Does Justice Look Like? Justice Resolution 001." August 7, 2020. https://assets1.cbsnewsstatic.com/i/cbslocal/wp-content/uploads/sites/15909630/2020/08/38th-Street-East-Chicago-Avenue-South-JUSTICE-RESOLUTION.pdf.

Merriam-Webster Online Dictionary. "Infrastructure." In *Merriam-Webster Online Dictionary*, January 2, 2018. https://www.merriam-webster.com/dictionary/infrastructure.

Metropolitan Council, Twin Cities metro area. "METRO D Line Construction Update-5/6/2022," May 6, 2022. https://content.govdelivery.com/accounts/MNORGMETC/bulletins/316b85b.

Miller, Johnny. "Roads to Nowhere: How Infrastructure Built on American Inequality." *The Guardian*, February 21, 2018. http://www.theguardian.com/cities/2018/feb/21/roads-nowhere-infrastructure-american-inequality.

Mills, Mara. "Body Meets World: Mara Mills Talks to Sara Hendren about Design and Disability." *Artforum*, November 11, 2020. https://www.artforum.com/books/mara-mills-talks-to-sara-hendren-about-design-and-disability-84333.

Mina, An Xiao, and Nicole Anand. "Elephant Problems: The Limitations of Expertise and the Possibilities of Multistakeholderism." Accessed May 17, 2021. https://anxiaomina.substack.com/p/elephant-problems-the-limitations.

Misharina, Anna, and Eleanor Betts. "The Embodied City: A Method for Multisensory Mapping." In *Capturing the Senses: Digital Methods for Sensory Archaeologies*, edited by

Giacomo Landeschi and Eleanor Betts, 237–264. Cham: Springer International Publishing, 2023.

Mittel, Jason. "A Few Thoughts on Improving the SCMS Conference." Just TV (blog). January 14, 2014. https://justtv.wordpress.com/2014/01/14/a-few-thoughts-on-improving-the-scms-conference/.

Modern Language Association. "The Future of Scholarly Publishing: MLA Ad Hoc Committee on the Future of Scholarly Publishing." *Profession*, 2 (2002): 172–186.

Modern Language Association. "Guidelines for Evaluating Publicly Engaged Humanities Scholarship in Language and Literature Programs. MLA Ad Hoc Committee on Valuing the Public Humanities." August 2022. https://www.mla.org/content/download/187094/file/Guidelines-Evaluating-Public-Humanities.pdf.

Montoya, Robert, and Gregory Leazer. "Limits of 'Infrastructure.'' Panel Description for the Annual Meeting of the Society for Social Studies of Science (4S), Boston, August 30–September 2, 2017. https://4sonline.org/docs/4S17_program_abstracts.pdf.

Morgan, Clara. "Higher Education Funding and Policy Trade-Offs: The AUCC and Federal Research in the Chrétien-Martin Era." In *Research and Innovation Policy: Changing Federal Government-University Relations*, edited by G. Bruce Doern and Christopher Stoney, 59–86. Toronto: University of Toronto Press, 2009.

Morris, Carol, Minna Kaljonen, Kadri Aavik, Bálint Balázs, Matthew Cole, Ben Coles, Sophia Efstathiou, Tracey Fallon, Mike Foden, Eva Haifa Giraud, Mike Goodman, Eleanor Hadley Kershaw, Richard Helliwell, Pru Hobson-West, Matti Häyry, Piia Jallinoja, Mat Jones, Taija Kaarlenkaski, Maarit Laihonen, Anu Lähteenmäki-Uutela, Saara Kupsala, Annika Lonkila, Lydia Martens, Renelle McGlacken, Josephine Mylan, Mari Niva, Emma Roe, Richard Twine, Markus Vinnari, and Richard White. "Priorities for Social Science and Humanities Research on the Challenges of Moving beyond Animal-Based Food Systems." *Humanities and Social Sciences Communications* 8 (February 3, 2021): 1–12.

Morrison, Romi Ron. "Gaps between the Digits: On the Fleshy Unknowns of the HUMAN." *Information Design Journal* 25, no. 1 (2019): 56–70.

Moteff, John, Claudia Copeland, and John Fischer. "Critical Infrastructures: What Makes an Infrastructure Critical?" Congressional Research Service, the Library of Congress, Washington, DC, 2003.

Mukerji, Chandra. *Impossible Engineering: Technology and Territoriality on the Canal Du Midi*. Princeton, NJ: Princeton University Press, 2009.

Mukerji, Chandra. "Me? A Digital Humanist?" In *Between Humanities and the Digital*, edited by Patrik Svensson and David Theo Goldberg, 41–53. Cambridge, MA: MIT Press, 2015.

Mullen, Mary L. "Public Humanities' (Victorian) Culture Problem." *Cultural Studies* 30, no. 2 (2016): 183–204.

National Council on Public Works Improvement. *Fragile Foundations: A Report on America's Public Works. Final Report to the President and the Congress.* Washington, DC: National Council on Public Works Improvement, 1988.

National Endowment for the Humanities. "About the National Endowment for the Humanities." December 8, 2017. https://www.neh.gov/about.

National Endowment for the Humanities. "Digging into Data Challenge." Accessed March 2, 2021. https://diggingintodata.org/about.

National Endowment for the Humanities. "NEH Chairman William D. Adams Statement on the Proposed Elimination of NEH in FY18 Budget | National Endowment for the Humanities." March 15, 2017. https://www.neh.gov/news/press-release/2017-03-16.

National Endowment for the Humanities. "Strategic Plan, Fiscal Year 2013–Fiscal Year 2017." n.d. Accessed September 20, 2017. http://web.archive.org/web/20171209010328/https://www.neh.gov/about/legal/strategic-plan.

National Humanities Center. "Fellows by the Numbers." National Humanities Center. Accessed February 13, 2021. https://nationalhumanitiescenter.org/fellows-by-the-numbers/.

National Science Foundation. "FY 2024 Budget Request to Congress (NSF)." March 13, 2023. https://nsf-gov-resources.nsf.gov/2023-08/NSF%20FY24%20CJ_Entire%20Rollup_web_%28ERRATA%20v4%29.pdf?VersionId=O1Um6rPm6xnLj80uA05Dm7IwxHUqScDO.

NATO. "Aspects of NATO Series 1—No. 12. Infrastructure and Logistics—NATO Archives Online." 1982. https://archives.nato.int/aspects-of-nato-series-1-n-12-infrastructure-and-logistics.

Nguyen, Jane (@theglowingstars). "This is for Joe whose heart medication was thrown out in a sweep, Chris whose mother's ashes was destroyed in a sweep, Khan who begged to keep his bedding, Janet Garcia who lost her cleaning supplies for work, including a cherished vacuum cleaner she named 'baby shark.'" Twitter, April 14, 2020.

Noble, Safiya Umoja. *Algorithms of Oppression: How Search Engines Reinforce Racism.* New York: NYU Press, 2018.

Noble, Safiya U., and Sarah T. Roberts. "Out of the Black Box." *EDUCAUSE Review*, March 13, 2017.

Norberg, Erik. *Mellan tiden och evigheten: Riksarkivet 1846–1991.* Stockholm: Riksarkivet, 2007.

Nordin, Jonas. "Plikten bakom allt: Pliktleveranshanteringen vid det kungliga biblioteket under 1700-talet." In *Information som problem. Medieanalytiska texter från medeltid till framtid*, edited by Otfried Czaika, Jonas Nordin, and Pelle Snickars, 10–43. Stockholm: Kungliga biblioteket, 2014.

North Atlantic Treaty Organization, Infrastructure Committee. *50 Years of Infrastructure: Nato Security Investment Programme is the Sharing of Roles, Risks, Responsibilities, Costs and Benefits*. Washington, DC: NATO Graphics Studio, 2001.

Nye, David E. *American Technological Sublime*. Cambridge, MA: MIT Press, 1994.

Occidental College. "Food Studies." Food Studies. Accessed July 20, 2022. https://www.oxy.edu/academics/areas-study/food-studies.

OECD Global Science Forum. "Report on Roadmapping of Large Research Infrastructures." 2008. https://web-archive.oecd.org/2012-06-14/99395-47057832.pdf.

Ohlin, Bertil. Motion 1961:363 Andra kammaren—Riksdagen, Pub. L. No. 1961:363 (1961). https://www.riksdagen.se/sv/dokument-och-lagar/dokument/motion/motioner-i-andra-kammaren-nr-363-ar-1961_em2o363/html/.

Olofsson, Jennie. "Gender, Class and Altered Soundscapes: Following the Implementation of a Robotic Welding System." *Ethnography* 19, no. 3 (2018): 379–395.

Oman, Georgia. *Higher Education and the Gendering of Space in England and Wales, 1869–1909*. Cham: Palgrave Macmillan, 2023.

Open Society University Network. "Experimental Humanities Collaborative Network (EHCN) | Open Society University Network." Accessed May 13, 2021. https://opensocietyuniversitynetwork.org/education/curricula/ehcn/.

Östh Gustafsson, Hampus. *Folkhemmets styvbarn: Humanioras legitimitet i svensk kunskapspolitik 1935–1980*. Göteborg: Daidalos, 2020.

Östh Gustafsson, Hampus. "Gadfly or Guide of Souls? The Challenge of Democracy to the Twentieth-Century Humanities." In *The Humanities and the Modern Politics of Knowledge: The Impact and Organization of the Humanities in Sweden, 1850–2020*, edited by Anders Ekström and Hampus Östh Gustafsson, 131–154. Amsterdam: Amsterdam University Press, 2022.

Östling, Johan, Anton Jansson, and Ragni Svensson Stringberg, eds. *Humaniora i välfärdssamhället: kunskapshistorier om efterkrigstiden*. Göteborg: Makadam förlag (Kriterium), 2023.

Östling, Johan, Katarina Bernhardsson, Jenny Björkman, and Jesper Olsson. "Kunskapens nya rörelser: Framtidens humanistiska och samhällsvetenskapliga samverkan." Kungliga Vitterhets-historie-och antikvitetsakademien, 2016. https://lucris.lub.lu.se/ws/portalfiles/portal/19770574/Kunskapens_nya_r_relser.pdf.

Oxford English Dictionary. "Infrastructure." In *OED Online*. Oxford University Press. Accessed October 21, 2020. http://www.oed.com/view/Entry/95624.

Padberg, Britta. "The Global Diversity of Institutes for Advanced Study." *Sociologica* 14, no. 1 (2020): 119–161.

Palm, John. "Professor hoppade av jämställdhetsprojekt på Chalmers – 'Inga förändringar på djupet.'" *Ny Teknik*, September 10, 2023.

Palmieri, Kristine. "The Forgotten Seminar: Friedrich Creuzer and Classical Philology at the University of Heidelberg, 1800–1830." *History of Humanities* 8, no. 1 (2023): 69–97.

Paoletti, Jo B., Mary Corbin Sies, and Virginia Jenkins. "Using a Virtual Museum for Collaborative Teaching, Research, and Service." In *Electronic Collaboration in the Humanities: Issues and Options*, edited by James A. Inman, Cheryl Reed, and Peter Sands, 151–165. New York: Routledge, 2003.

Papon, Pierre. "European Scientific Cooperation and Research Infrastructures: Past Tendencies and Future Prospects." *Minerva* 42, no. 1 (2004): 61–76.

Pargman, Daniel, and Elina Eriksson. "'It's Not Fair!': Making Students Engage in Sustainability." In *Proceedings of the EESD13 Conference*, Cambridge, UK, September 22–25, 2013.

Parham, Marisa. "Ninety-Nine Problems: Assessment, Inclusion, and Other Old-New Problems." *American Quarterly* 70, no. 3 (2018): 677–684.

Parks, Lisa. "'Stuff You Can Kick': Toward a Theory of Media Infrastructures." In *Between Humanities and the Digital*, edited by Patrik Svensson and David Theo Goldberg, 355–373. Cambridge, MA: MIT Press, 2015.

Parks, Lisa. "Technostruggles and the Satellite Dish: A Populist Approach to Infrastructure." In *Cultural Technologies: The Shaping of Culture in Media and Society*, edited by Göran Bolin, 64–84. London: Routledge, 2012.

Parks, Lisa, and Nicole Starosielski. "Introduction." In *Signal Traffic: Critical Studies of Media Infrastructures*, edited by Lisa Parks and Nicole Starosielski, 1–27. Urbana: University of Illinois Press, 2015.

Parks, Lisa, and Nicole Starosielski, eds. *Signal Traffic: Critical Studies of Media Infrastructures*. Urbana: University of Illinois Press, 2015.

Pasek, Anne. "Fixing Carbon: Mediating Matter in a Warming World." New York University, 2019. https://www.proquest.com/openview/467b86fbcc43db903dee71f7b7f097f6/1?pq-origsite=gscholar&cbl=18750&diss=y.

Pasek, Anne. "Low-Carbon Research: Building a Greener and More Inclusive Academy." *Engaging Science, Technology, and Society* 6 (2020): 34–38.

Pasek, Anne. "On Being Anxious about Digital Carbon Emissions." *Social Media + Society* 9, no. 2 (April 1, 2023).

Pasek, Anne, Emily Roehl, and Caleb Wellum. "Making and Meeting Online: A White Paper on E-Conferences, Workshops, and Other Experiments in Low-Carbon Research Exchange." *Energy Humanities*, 2020. https://hcommons.org/deposits/item/hc:41217.

Patterson, G. E. "George Floyd and a Community of Care." *Places Journal*, January 2021. https://placesjournal.org/article/george-floyd-and-a-community/.

Patton, Carl V. "Infrastructure Decay in the United States." *Built Environment (1978–)* 10, no. 4 (1984): 231–244.

Pedersen, David Budtz, Jonas Følsgaard Grønvad, and Rolf Hvidtfeldt. "Methods for Mapping the Impact of Social Sciences and Humanities—A Literature Review." *Research Evaluation* 29, no. 1 (2020): 4–21.

Pelizza, Annalisa. "Developing the Vectorial Glance: Infrastructural Inversion for the New Agenda on Government Information Systems." *Science, Technology, and Human Values* 41, no. 2 (2016): 298–321.

Peters, John Durham. *The Marvelous Clouds: Toward a Philosophy of Elemental Media*. Chicago: University of Chicago Press, 2015.

Peters, John Durham, and Theodore Kim. "Critical Humanities Could Bring Good Trouble to AI." *Yale Daily News*, January 24, 2024. https://yaledailynews.com/blog/2024/01/24/peters-kim-critical-humanities-could-bring-good-trouble-to-ai/.

Peters, Kimberley, and Jennifer Turner. "Unlock the Volume: Towards a Politics of Capacity." *Antipode* 50, no. 4 (2018): 1037–1056.

Phaal, Robert, Clare J. P. Farrukh, and David R. Probert. "Technology Roadmapping—A Planning Framework for Evolution and Revolution." *Technological Forecasting and Social Change*, Roadmapping: From Sustainable to Disruptive Technologies, 71, no. 1 (2004): 5–26.

Plantin, Jean-Christophe, and Aswin Punathambekar. "Digital Media Infrastructures: Pipes, Platforms, and Politics." *Media, Culture and Society* 41, no. 2 (2019): 163–174.

Plantin, Jean-Christophe, Carl Lagoze, Paul N. Edwards, and Christian Sandvig. "Infrastructure Studies Meet Platform Studies in the Age of Google and Facebook." *New Media and Society* 20, no. 1 (2018): 293–310.

Plotnikoff, Elena. "Activists or Active Threats?: How the State Securitization of Critical Infrastructure Impacts Environmental and Indigenous Activists in Canada and the United States." University of British Columbia, 2020. https://open.library.ubc.ca/soa/cIRcle/collections/ubctheses/24/items/1.0394048.

Poirier, Lindsay. "Devious Design: Digital Infrastructure Challenges for Experimental Ethnography." *Design Issues* 33, no. 2 (April 2017): 70–83.

Posner, Miriam. "Breakpoints and Black Boxes: Information in Global Supply Chains." *Postmodern Culture* 31, no. 3 (2021).

Powell, Alison B. *Undoing Optimization: Civic Action in Smart Cities*. New Haven, CT: Yale University Press, 2021.

Pratt, Mary Louise. "Arts of the Contact Zone." *Profession*, 1991, 33–40.

Presner, Todd. "The Ethics of the Algorithm: Close and Distant Listening to the Shoah Foundation Visual History Archive." In *Probing the Ethics of Holocaust Culture*, edited by Claudio Fogu, Wulf Kansteiner, and Todd Presner, 175–202. Cambridge, MA: Harvard University Press, 2016.

Presner, Todd, Peter Lunenfeld, and Jeffrey Schnapp. "The Digital Humanities Manifesto 2.0." 2009, 15. https://www.humanitiesblast.com/manifesto/Manifesto_V2.pdf.

Public Safety Canada. "Canada's Critical Infrastructure." October 26, 2022. https://www.publicsafety.gc.ca/cnt/ntnl-scrt/crtcl-nfrstrctr/cci-iec-en.aspx.

Qadri, Rida [@qadrida]. "Excited to read this piece by the always incisive @tetisheri. Critical tech scholars must turn the critical lens inwards to investigate our own research practices, methods and the power asymmetries we perpetuate." Twitter, July 12, 2021.

Ramakrishnan, Kavita, Kathleen O'Reilly, and Jessica Budds. "The Temporal Fragility of Infrastructure: Theorizing Decay, Maintenance, and Repair." *Environment and Planning E: Nature and Space* 4, no. 3 (2021): 674–695.

Ratto, Matt. "Critical Making: Conceptual and Material Studies in Technology and Social Life." *The Information Society* 27, no. 4 (2011): 252–260.

Ratto, Matt. "Epistemic Commitments, Virtual Reality, and Archaeological Representation." In *Technology and Methodology for Archaeological Practice: Proceedings of the XV UISPP World Congress (Lisbon, 4–9 September 2006)*, edited by Alexandra Velho and Hans Kamermans. Oxford: Archaeopress, 2009.

Ratto, Matt. "Greater and Less than Human? AI and Behavioral Intervention Bots (SRI Seminar Series)." Schwartz Reisman Institute, University of Toronto, August 19, 2021. https://www.youtube.com/watch?v=MmYcV_CEcnQ.

Ratto, Matt. "Making at the End of Nature." *Interactions* 23, no. 5 (2016): 26–35.

Ratto, Matt. "Not Just Guns but Bullets, Too: 'Deconstructive' and 'Constructive' Making within the Digital Humanities." In *Debates in the Digital Humanities 2019*, edited by Matthew K. Gold and Lauren F. Klein, 307–318. Minneapolis: University of Minnesota Press, 2019.

Ratto, Matt, and Steven J. Jackson. "Reopening, Repetition and Resetting: HCI and the Method of Hope." *Proceedings of the ACM on Human-Computer Interaction* 7, no. CSCW2 (October 2023): 356:1–356:30.

Rees, Jonathan. "The Many Ways That Water Froze: A Taxonomy of Ice in Nineteenth- and Early Twentieth-Century America." In *Ice Humanities: Living, Working, and Thinking in a Melting World*, edited by Klaus Dodds and Sverker Sörlin, 154–167. Manchester: Manchester University Press, 2022.

Regeringskansliet (Swedish Government). "Nya styrelser för 30 universitet och högskolor." Regeringskansliet. Regeringen och Regeringskansliet, April 27, 2023. https://www.regeringen.se/pressmeddelanden/2023/04/nya-styrelser-for-30-universitet-och-hogskolor/.

Regeringskansliet (Swedish Government). "Stora satsningar på forskning i årets budget." Regeringskansliet. Regeringen och Regeringskansliet, September 15, 2020. https://www.regeringen.se/artiklar/2020/09/stora-satsningar-pa-forskning-i-arets-budget/.

Reid, Julian. "Conclusion: The Biopolitics of Critical Infrastructure Protection." In *Securing "the Homeland": Critical Infrastructure, Risk and (In)Security*, edited by Myriam Cavelty Dunn and Kristian Søby Kristensen, 176–183. New York: Routledge, 2008.

Rentschler, Carrie. "The Eavesdropper and Onlooker as Proximate Agents of Social Change." *Media Theory* 5, no. 2 (2021): 261–288.

Reslow, Patrick, Robert Stenkvist, Michael Rubbestad, and Jörgen Grubb. "SD: Stoppa Genusindoktrinering (Opinion Piece, Sweden Democrats)." *Altinget*, April 8, 2019.

Ribes, David, and Steven J. Jackson. "Data Bite Man: The Work of Sustaining a Long-Term Study." In *"Raw Data" Is an Oxymoron*, edited by Lisa Gitelman, 147–166. Cambridge, MA: MIT Press, 2013.

Ridder-Symoens, Hilde de, ed. *A History of the University in Europe: Volume 2, Universities in Early Modern Europe*. Cambridge: Cambridge University Press, 1996.

Riklund, Katrine. "Del 1:2 Forskningsinfrastruktur." 2020. Dialogseminarium för forskningsfinansiärer och lärosäten (SUHF), Stockholm, January 24, 2020. https://suhf.se/app/uploads/2020/01/Forskningsinfrastruktur-URFI-Katrine-Riklund-UmU-Dialogsem.-200124.pdf.

Riksarkivet. "Årsredovisning 2019." 2020. https://riksarkivet.se/Media/pdf-filer/ra_arsredovisning__2019_klar.pdf.

Riksarkivet. "Årsredovisning 2020." 2021. https://riksarkivet.se/Media/pdf-filer/ra_arsredovisning_2020_klar.pdf.

Riksarkivet. "Årsredovisning 2021." 2022. https://riksarkivet.se/Sve/Remisser/Dokument/RA_%C3%85rsredovisning_2021.pdf.

Riksarkivet. "Årsredovisning 2022." 2023. https://riksarkivet.se/Sve/Remisser/Dokument/%C3%85rsredovisning%202022.pdf.

Riksbankens Jubileumsfond. "Redogörelse till Riksdagen 1996/97:RJ." 1997. https://data.riksdagen.se/fil/1B47D450-CBCC-4368-80E9-C4FFBCD6E3C3.

Riksdagsförvaltningen (Swedish Government). "Dokument & lagar." Accessed April 10, 2023. https://www.riksdagen.se/sv/dokument-lagar/.

Riley, Shannon Rose. "Lab/Studio." In *Mapping Landscapes for Performance as Research: Scholarly Acts and Creative Cartographies*, edited by Shannon Rose Riley and Lynette Hunter, 137–141. London: Palgrave Macmillan UK, 2009.

Risam, Roopika. *New Digital Worlds: Postcolonial Digital Humanities in Theory, Praxis, and Pedagogy*. Evanston, IL: Northwestern University Press, 2018.

Risam, Roopika, and Alex Gil. "Introduction: The Questions of Minimal Computing." *Digital Humanities Quarterly* 16, no. 2 (2022).

Risbjerg Eskildsen, Kasper. "The Humanities in the Vocational University: On the Unity of Teaching and Research." In *Writing the History of the Humanities: Questions, Themes, and Approaches*, edited by Herman Paul, 267–285. London: Bloomsbury Academic, 2023.

Risbjerg Eskildsen, Kasper. "Virtues of History: Exercises, Seminars, and the Emergence of the German Historical Discipline, 1830–1900." In *History of Universities: Volume XXXIV/1: A Global History of Research Education: Disciplines, Institutions, and Nations,*

1840–1950, edited by Ku-ming (Kevin) Chang and Alan Rocke, 27–40. Oxford: Oxford University Press, 2021.

Robles-Anderson, Erica. "The Crystal Cathedral: Architecture for Mediated Congregation." *Public Culture* 24 (2012): 577–599.

Robles-Anderson, Erica, and Patrik Svensson. "'One Damn Slide after Another': PowerPoint at Every Occasion for Speech." *Computational Culture*, 5 (2016).

Rosner, Daniela K. *Critical Fabulations: Reworking the Methods and Margins of Design.* Cambridge, MA: MIT Press, 2018.

Rossitto, Chiara, Airi Lampinen, Susanne Bødker, Ann Light, Ketie Berns, and Julie Hui. "Reconsidering Scale and Scaling in CSCW Research." In *Companion Publication of the 2020 Conference on Computer Supported Cooperative Work and Social Computing*, edited by Matthew Bietz and Andrea Wiggins, 493–501. CSCW '20 Companion. New York: Association for Computing Machinery, 2020.

Rothstein, Bo. "Myndigheter att lita på. Den svenska demokratins grundbult." In *Demokratins framtid*, edited by Katarina Barrling and Sören Holmberg, 211–236. Stockholm: Sveriges riksdag, 2018.

Royal Holloway Centre for the GeoHumanities. "GeoHumanities Forum (About Page)." Accessed May 30, 2021.

Runting, Helen, Rutger Sjögrim, and Fredrik Torisson. "Pop Theory: The Architecture of Late Night Shopping." *Architecture and Culture* 5, no. 3 (2017): 513–524.

Salö, Linus, Sverker Sörlin, and Mats Benner. "En inledning: Humanvetenskapernas kunskapspolitik." In *Humanvetenskapernas verkningar*, edited by Linus Salö, 9–34. Stockholm: Dialogos Förlag, 2021.

Samorani, Michele, Shannon L. Harris, Linda Goler Blount, Haibing Lu, and Michael A. Santoro. "Overbooked and Overlooked: Machine Learning and Racial Bias in Medical Appointment Scheduling." *Manufacturing and Service Operations Management* 24, no. 6 (2022): 2825–2842.

Sandvig, Christian. "The Internet as Infrastructure." In *The Oxford Handbook of Internet Studies*, edited by William H. Dutton, 86–106. Oxford: Oxford University Press, 2013.

Scagnetti, Gaia. "A Dialogical Model for Studio Critiques in Design Education." *The Design Journal* 20, sup1 (2017): S781–S791.

Schnapp, Jeffrey, and Michael Shanks. "Artereality (Rethinking Craft in a Knowledge Economy)." In *Art School (Propositions for the 21st Century)*, edited by Steven Henry Madoff, 141–157. Cambridge, MA: MIT Press, 2009.

School of Education and Information Studies, UCLA. "Johanna Drucker: The Future of Special Collections Libraries." February 12, 2020. Accessed January 10, 2021. http://web.archive.org/web/20201029111547/https://ampersand.gseis.ucla.edu/johanna-drucker-the-future-of-special-collections-libraries/.

Science|Business. “Lund Declaration: Let Grand Challenges Drive EU Research.” *Science|Business*, July 8, 2009.

Scott-Webber, Lennie. *In Sync: Environmental Behavior Research and the Design of Learning Spaces*. Ann Arbor, MI: Society for College and University Planning, 2004.

Sha, Xin Wei. “The Atelier-Lab as a Transversal Machine.” *Revue Française d'Études Américaines* 128, no. 2 (2011): 62–78.

Sha, Xin Wei, and Patrik Svensson. “ASU-SCAS Workshop Building Humanistic and Humane Capacity, May 5, 2022.” 2022. https://docs.google.com/document/d/1JaJwYFPzf3mEfl8FzQrOoXH7EKQh230wiD5PNxe5BRw/.

SHAPE-ID: Shaping Interdisciplinary Practices in Europe. “Christian Pohl Discusses the Role of the Humanities in Addressing Societal Challenges (Video Recording).” n.d. https://www.shapeidtoolkit.eu/#section_2.

Shapin, Steven, and Simon Schaffer. *Leviathan and the Air-Pump: Hobbes, Boyle, and the Experimental Life*. Princeton, NJ: Princeton University Press, 1985.

Simone, AbdouMaliq. “People as Infrastructure: Intersecting Fragments in Johannesburg.” *Public Culture* 16, no. 3 (2004): 407–429.

Siskin, Clifford. *System: The Shaping of Modern Knowledge*. Cambridge, MA: MIT Press, 2016.

Skelton, Alan. “Conferences, Conferences, Conferences?” *Teaching in Higher Education* 2, no. 1 (1997): 69–72.

Small, Helen. *The Value of the Humanities*. Oxford: Oxford University Press, 2013.

Small, Mario L. “Departmental Conditions and the Emergence of New Disciplines: Two Cases in the Legitimation of African-American Studies.” *Theory and Society* 28, no. 5 (1999): 659–707.

Smarr, Larry, John Kogut, David Kuck, Robert Wilhelmson, Peter Wolynes, Karl Hess, Thomas Hanratty, and Robert McMeeking. “A Center for Scientific and Engineering Supercomputing (NSF Proposal).” 1983. https://lsmarr.net/a-center-for-scientific-and-engineering-supercomputing/.

Smith, Bonnie G. “Gender and the Practices of Scientific History: The Seminar and Archival Research in the Nineteenth Century.” *The American Historical Review* 100, no. 4 (1995): 1150–1176.

Smith, Martha Nell. “The Human Touch Software of the Highest Order: Revisiting Editing as Interpretation.” *Textual Cultures: Texts, Contexts, Interpretation* 2, no. 1 (2007): 1–15.

Smithies, James. *The Digital Humanities and the Digital Modern*. London: Palgrave Macmillan UK, 2017.

Snickars, Pelle. *Kulturarvets mediehistoria: dokumentation och representation 1750–1950*. Lund: Mediehistoriskt arkiv, 2020.

Social Sciences and Humanities Research Council (Canada). "Community Consultation on Infrastructure in the Social Sciences and Humanities." May 2009. https://www.sshrc-crsh.gc.ca/about-au_sujet/publications/infrastructure_CFI_e.pdf.

Society for Cinema and Media Studies. "2022 SCMS Conference Update (Video)." January 10, 2022. https://myemail.constantcontact.com/2022-SCMS-Conference-Update.html?soid=1121577869045&aid=fe62xkNlL_M.

Society for Cinema and Media Studies. "Past SCMS Conferences." Accessed August 15, 2024. https://www.cmstudies.org/page/past_conferences.

Society for Cinema and Media Studies. "SCMS 2019 Conference Program." 2019. https://cdn.ymaws.com/www.cmstudies.org/resource/resmgr/2019_conference/SCMS2019INTFP-webversion.pdf.

Society for Cinema and Media Studies. "SCMS 2020 Cancellation Official Statement." 2020. https://www.cmstudies.org/page/officialstatement.

Society for Cinema and Media Studies. "SCMS Annual Conference." Accessed October 2, 2019. https://www.cmstudies.org/page/conference.

Society for Cinema and Media Studies. "SCMS Members Meeting 2022, 2022 Virtual Conference (Presentation)." March 12, 2022. https://cdn.ymaws.com/www.cmstudies.org/resource/resmgr/2022_conference/SCMS_MembersMeeting_2022__1_.pdf.

Society for Cinema and Media Studies. "SCMS Organizational History." Accessed February 8, 2024. https://www.cmstudies.org/page/org_history.

Society for Cinema and Media Studies. "Transcript: SCMS 2022 Conference Decision (February 3)." Google Docs, February 3, 2022. https://docs.google.com/document/d/1AGAQ5w2JkUlYi7fCBbzhfg4Hi1ZeWAjMsehQrh1epLo/.

Sörlin, Sverker. "Humanities of Transformation: From Crisis and Critique towards the Emerging Integrative Humanities." *Research Evaluation* 27, no. 4 (October 2018): 287–297.

Sörlin, Sverker. "Så blev Stockholm en världsstad i kampen mot miljöförstörelsen." *Dagens Nyheter*, May 30, 2022.

Sörlin, Sverker. "The Changing Nature of Environmental Expertise." *Eurozine*, November 19, 2013. https://www.eurozine.com/the-changing-nature-of-environmental-expertise/.

Sörlin, Sverker, and Eric Paglia. "Stockholm and 1972—Capital of Environmental Memory." *Kungl. Skogs- Och Lantbruksakademiens Tidskrift* 161, no. 1 (2022): 89–95.

Spee, Andreas Paul, and Paula Jarzabkowski. "Strategy Tools as Boundary Objects." *Strategic Organization* 7, no. 2 (2009): 223–232.

Spice, Anne. "Fighting Invasive Infrastructures: Indigenous Relations against Pipelines." *Environment and Society* 9, no. 1 (2018): 40–56.

Stamm, Julia. "Interdisciplinarity Put to Test: Science Policy Rhetoric vs Scientific Practice—the Case of Integrating the Social Sciences and Humanities in Horizon 2020." In *Handbook on Science and Public Policy*, edited by Dagmar Simon, Stefan Kuhlmann, Julia Stamm, and Weert Canzler, 376–399. Cheltenham, UK: Edward Elgar, 2019.

Stanford Humanities Center. "2015–16 Annual Report: Dialogue." Stanford University, 2016. https://shc.stanford.edu/sites/default/files/2022-07/SHC_16AnnualReport_Web.pdf.

Stanford University. "Introducing the Stanford Doerr School of Sustainability." *Stanford News* (blog). May 4, 2022. https://news.stanford.edu/2022/05/04/stanford-doerr-school-sustainability-universitys-first-new-school-70-years-will-accelerate-solutions-global-climate-crisis/.

Star, Susan Leigh. "The Ethnography of Infrastructure." *American Behavioral Scientist* 43, no. 3 (1999): 377–391.

Star, Susan Leigh, and James R. Griesemer. "Institutional Ecology, 'Translations' and Boundary Objects: Amateurs and Professionals in Berkeley's Museum of Vertebrate Zoology, 1907–39." *Social Studies of Science* 19, no. 3 (1989): 387–420.

Star, Susan Leigh, and Karen Ruhleder. "Steps Toward an Ecology of Infrastructure: Design and Access for Large Information Spaces." *Information Systems Research* 7, no. 1 (1996): 111–134.

Steinhart, Eric, and Eva Kittay. "Generating Metaphors from Networks: A Formal Interpretation of the Semantic Field Theory of Metaphor." In *Aspects of Metaphor*, edited by Jaakko Hintikka, Synthese Library, Volume 238, 41–94. Dordrecht: Springer, 1994.

Sterne, Jonathan. "At the Interface (Course Curriculum, Spring 2023)." McGill University, 2023. https://sterneworks.org/COMS%20648-W23-4JanWEB.pdf.

Stewart, Kathleen. *Ordinary Affects*. Durham, NC: Duke University Press, 2007.

Stirling, Andy. "Direction, Distribution and Diversity! Pluralising Progress in Innovation, Sustainability and Development." STEPS Working Paper, STEPS Centre, Birmingham, UK, 2009.

Stockholm University. "Forskningsinfrastruktur." Forskningsinfrastruktur. Accessed November 18, 2019. https://www.su.se/forskning/forskningsinfrastruktur.

Stone, Allucquére Rosanne. "On Being Trans, and under the Radar: Tales from the ACT-Lab," *Switch* 20 (2005). https://switch.sjsu.edu/archive/v20/Articles/sandystone/index.html.

Strategic Concepts in Organizing and Policy Education (SCOPE). "Between the 110 and the 405: Environmental Injustice in South Los Angeles." SCOPE Blog (blog). November 27, 2017. https://scopela.org/between-the-110-and-the-405-environmental-injustice-in-south-los-angeles/.

Strelka Institute/Институт Стрелка, and Benjamin Bratton. "The Terraforming 2022. New Program Presentation (Video)." September 13, 2021. https://www.youtube.com/watch?v=v4NZS1ck02o.

Striphas, Ted. "Introduction the Long March: Cultural Studies and Its Institutionalization." *Cultural Studies* 12, no. 4 (1998): 453–475.

Students4Students. "College Homelessness: By the Numbers." Students4Students, n.d. Accessed August 1, 2020. http://web.archive.org/web/20201112022814/https://www.s4sla.org/student-homelessness-1.

Sumathipala, Athula. "Research Infrastructure." In *Essentials of Global Mental Health*, edited by Samuel O. Okpaku, 407–415. Cambridge: Cambridge University Press, 2014.

Sundén, Jenny. "On Trans-, Glitch, and Gender as Machinery of Failure." *First Monday* 20, no. 4 (2015).

Svensson, Daniel, Sverker Sörlin, and Nina Wormbs. "The Movement Heritage: Scale, Place, and Pathscapes in Anthropocene Tourism." In *Tourism and the Anthropocene*, edited by Martin Gren and Edward Huijbens, 131–151. London: Routledge, 2015.

Svensson, Ingrid. "Kapitel 4: Nationalbiblioteket i Humlegården, 1800-talet." In *Kungliga biblioteket i ord och bild*, edited by Ingrid Svensson. Stockholm: Kungliga biblioteket, 2018.

Svensson, Patrik. *Big Digital Humanities: Imagining a Meeting Place for the Humanities and the Digital*. Ann Arbor, MI: University of Michigan Press, 2016.

Svensson, Patrik. "Contemporary and Future Spaces for Media Studies and Digital Humanities." In *The Routledge Companion to Media Studies and Digital Humanities*, edited by Jentery Sayers, 152–161. New York: Routledge, 2018.

Svensson, Patrik. "Critical Visualization Event at the Graduate Center, CUNY on June 9, 2016." 2016. https://patriksv.com/curatorship/#critviz.

Svensson, Patrik. "Envisioning the Digital Humanities." *Digital Humanities Quarterly* 6, no. 1 (2012).

Svensson, Patrik. "From Optical Fiber to Conceptual Cyberinfrastructure." *Digital Humanities Quarterly* 5, no. 1 (2011).

Svensson, Patrik, ed. "Humane Infrastructures Conversation #1: Inside and Outside: Shifting Socio-Technological Systems." 2021. https://drive.google.com/file/d/17CWAxC37vH-dStf8UtY9yE86z1TCcSG2/.

Svensson, Patrik, ed. "Humane Infrastructures Conversation #2: Changing Ourselves—Climate, Carbon and Concepts." 2021. Unpublished document.

Svensson, Patrik. "Humanistisk infrastruktur (main author: Svensson)." Stockholm: Riksbankens Jubileumsfond, 2019. https://www.rj.se/.

Svensson, Patrik. "Humanities Computing as Digital Humanities." *Digital Humanities Quarterly* 3, no. 3 (2009).

Svensson, Patrik. *Number and Countability in English Nouns: An Embodied Model*. Acta Universitatis Umensis. Uppsala: Swedish Science Press Uppsala, 1998.

Svensson, Patrik. "The Digital Humanities as a Humanities Project." *Arts and Humanities in Higher Education* 11, no. 1–2 (2012): 42–60.

Svensson, Patrik. "The Humanistiscope: Exploring the Situatedness of Humanities Infrastructure." In *Between Humanities and the Digital*, edited by Patrik Svensson and David Theo Goldberg, 337–354. Cambridge, MA: MIT Press, 2015.

Svensson, Patrik. "Thoughts on Making Academic Events," October 15, 2019. Accessed August 13, 2024. https://patriksv.com/2019/10/some-thoughts-on-making-academic-events/.

Svensson, Patrik. "Universitetet som uttryck." In *Universitetet som medium*, edited by Matts Lindström and Adam Wickberg Månsson, 93–107. Lund: Mediehistoria, Lunds universitet, 2015.

Svensson, Patrik. "Virtual Worlds as Arenas for Language Learning." In *Language Learning Online: Towards Best Practice*, edited by Felix Uschi, 123–142. Lisse: Swets and Zeitlinger, 2003.

Sveriges riksdag. Förordning (2008:1421) med instruktion för Kungl. biblioteket (fram till SFS 2018:15). 2008/2018. https://www.riksdagen.se/sv/dokument-och-lagar/dokument/svensk-forfattningssamling/forordning-20081421-med-instruktion-for-kungl_sfs-2008-1421/.

Sveriges unga akademi, and Sällskapet riksdagsledamöter och forskare (RIFO). "Framtiden för svensk forskningsinfrastruktur (seminar)." December 2, 2020. https://sverigesungaakademi.se/event/framtiden-for-svensk-forskningsinfrastruktur/.

Sveriges universitets- och högskoleförbund (SUHF). "Dialogseminarium för forskningsfinansiärer och lärosäten (notes from seminar)." January 13, 2017. https://suhf.se/in-english/.

Sveriges universitets- och högskoleförbund (SUHF). "Till regeringen med anledning av beslut om förkortad mandatperiod för universitets- och högskolestyrelserna." April 28, 2023. https://suhf.se/app/uploads/2023/04/Till-regeringen-Utb-Land-Skrivelse-angaende-styrelserna-SUHF-SU-850-0024-23-REV.pdf.

Swedish Collegium for Advanced Study. "SCAS: The Natural Sciences Programme." Accessed February 11, 2021. http://www.swedishcollegium.se/subfolders/Fellowships/Natural_Sciences.html.

Synthesis Center, Arizona State University. "Alter-Eco: Alternate Economies and Ecologies." Synthesis. Accessed March 17, 2024. https://synthesiscenter.net/projects/alter-economies/.

Synthesis Center, Arizona State University. "Prototyping Social Forms." Prototyping Social Forms. Accessed August 28, 2022. https://prototyping-social-forms.weebly.com/.

Szeman, Imre, and Dominic Boyer. "Introduction: On the Energy Humanities." In *Energy Humanities: An Anthology*, edited by Imre Szeman and Dominic Boyer, 1–13. Baltimore: John Hopkins University Press, 2017.

The Daily Californian. "UCPath Has Left Employees Facing Financial Instability." *Daily Californian*, July 25, 2019.

The European Spallation Source. "About ESS." Accessed June 22, 2018. https://europeanspallationsource.se/about.

The Franke Institute for the Humanities, the University of Chicago. "The Franke Institute for the Humanities." Accessed February 12, 2021. https://www.franke.uchicago.edu.

The Institute for Advanced Study Princeton. "The Institute for Advanced Study." May 4, 2019. https://www.ias.edu/press/mission.

The International Union of Pure and Applied Physics. "UL Opara: 'With Research Infrastructure, Young Africans Will Perform Wonders.'" International Year of Basic Sciences for Development 2022–2023 (with UNESCO). Interview with Umezuruike Linus Opara., May 27, 2023. https://www.iybssd2022.org/en/ul-opara-with-research-infrastructure-young-africans-will-perform-wonders/.

The National Gallery, London. "Building the Picture | Antonello Da Messina | Saint Jerome in His Study." Accessed November 7, 2018. https://www.nationalgallery.org.uk/research/exhibition-catalogues/building-the-picture/entering-the-picture/antonello-da-messina-saint-jerome-in-his-study.

The Research Council of Norway. "Evaluation of KLIMAFORSK." 2021. https://www.forskningsradet.no/siteassets/publikasjoner/2021/evaluation-report-klimaforsk.pdf.

The Research Council of Norway. "Evaluation of the Humanities in Norway. Report from the Principal Evaluation Committee." June 2017. https://www.forskningsradet.no/siteassets/publikasjoner/1254027749230.pdf.

The Research Council of Norway. "MEATigation." Prosjektbanken—Meatigation. Accessed July 21, 2022. https://prosjektbanken.forskningsradet.no/project/FORISS/303698.

Thornton, Dora. *The Scholar in His Study: Ownership and Experience in Renaissance Italy*. New Haven, CT: Yale University Press, 1997.

Thylstrup, Nanna Bonde. *The Politics of Mass Digitization*. Cambridge, MA: MIT Press, 2019.

Tillväxtverket [@Tillväxtverket_konf]. "Nobelpris i sikte!Spännande ws med bl.a. @Naringsdep om framtiden för världsledande forskningsanläggningen ESS i Lund som @tillvaxtverket medfinansierar." Twitter, November 24, 2017.

Townsend, Robert B., and Norman Bradburn. "The State of the Humanities circa 2022." *Dædalus* 151, no. 3 (2022): 11–18.

Tracy, Karen. *Colloquium: Dilemmas of Academic Discourse*. Norwood, NJ: Ablex, 1997.

Travis, Charles, Deborah P. Dixon, Luke Bergmann, Robert Legg, and Arlene Crampsie, eds. *Routledge Handbook of the Digital Environmental Humanities*. New York: Routledge, 2022.

Turner, Graeme. *British Cultural Studies: An Introduction*. 3rd ed. London: Routledge, 2003.

UC Humanities Research Institute, and San Diego Supercomputer Center at UC San Diego. "Summer Institute: Cyberinfrastructure." *Humanist* 20, no. 008 (May 11, 2006).

UC Irvine. "EVOKE Lab and Studio—• Theorize • Design • Make • Provoke • Evoke • Inspire." Accessed March 6, 2021. https://evoke.ics.uci.edu/.

UC Riverside. "2nd Annual Digital Humanities Infrastructure Symposium." UCR | Center for Ideas and Society. Accessed November 7, 2018. https://ideasandsociety.ucr.edu/news/2nd-annual-digital-humanities-infrastructure-symposium/.

UC Santa Barbara. "The American Presidency Project." https://www.presidency.ucsb.edu/.

UCLA. "Impact | UCLA 100." 2019. https://100.ucla.edu/impact.

UCLA. "Out of the Classroom, Onto the Streets | UCLA 100." 2019. https://100.ucla.edu/impact/out-of-the-classroom-onto-the-streets.

UCLA Humanities Division. "UCLA Experimental Humanities Workshop on 'Humane Infrastructures.'" Accessed May 13, 2021. https://humanities.ucla.edu/event/ucla-experimental-humanities-workshop-on-humane-infrastructures/.

UCLA Newsroom. "UCLA to Launch New Social Justice Curriculum with $5 Million Grant from Mellon Foundation." Accessed May 13, 2021. https://newsroom.ucla.edu/releases/social-justice-curriculum-mellon-foundation-grant.

Umeå University. "Forskningsinfrastruktur." Accessed November 18, 2019. https://www.umu.se/forskning/forskningsinfrastruktur/.

Umeå University. "Humlab." Accessed August 26, 2019. https://www.umu.se/en/humlab/.

Umeå University. "Instruktion för Wallenberg AI, Autonomous Systems and Software Program—Humanities and Society." October 29, 2019. https://www.umu.se/globalassets/fristaende-webbar/regelverk/beslutsstruktur-delegation-och-organisation/instruktion-for-wallenberg-ai-autonomous-systems-and-software-program--humanities-and-society-wasp-hs-fs-1.1-19.pdf-407-kb_tillganp.pdf.

Umeå University. "Universitetsledningens externa uppdrag." Accessed August 7, 2024. https://www.umu.se/om-umea-universitet/sa-ar-vi-organiserade/universitetsledningens-externa-uppdrag/.

University of Warwick. "Arts Faculty Facilities." Accessed March 3, 2024. https://warwick.ac.uk/fac/arts/aboutus/artsfacultyfacilities/#fab.

University of Wisconsin–Madison. "About Us | UW–Madison Center for the Humanities." Accessed August 17, 2022. https://humanities.wisc.edu/about-us.

University of Wisconsin–Madison. "History (Institute for Research in the Humanities)." Accessed August 10, 2024. https://irh.wisc.edu/history/.

University of Wisconsin–Madison. "Institute for Research in the Humanities." Accessed February 20, 2021. https://irh.wisc.edu/.

Uppsala University. "Infrastruktur." Infrastruktur. Accessed November 18, 2019. https://www.uu.se/forskning/infrastruktur/.

Utaberta, Nangkula, Badiossadat Hassanpour, Aisyah Nur Handryant, and Adi Irfan Che Ani. "Upgrading Education Architecture by Redefining Critique Session in Design Studio." *Procedia—Social and Behavioral Sciences* 102 (2013): 42–47.

Venkatesan, Soumhya, Laura Bear, Penny Harvey, Sian Lazar, Laura Rival, and AbdouMaliq Simone. "Attention to Infrastructure Offers a Welcome Reconfiguration of Anthropological Approaches to the Political." *Critique of Anthropology* 38, no. 1 (2018): 3–52.

Vetenskapsrådet. "Årsredovisning 2001." 2002. https://www.vr.se/analys/rapporter/vara-rapporter/2002-11-11-arsredovisning-2001.html.

Vetenskapsrådet. "Årsredovisning 2022." 2023. https://www.vr.se/analys/rapporter/vara-rapporter/2023-03-01-arsredovisning-2022.html.

Vetenskapsrådet. "Bilagor Forskningsöversikt 2019 humaniora och samhällsvetenskap." 2019. https://www.vr.se/analys/rapporter/vara-rapporter/2019-01-29-forskningsoversikt-2019-humaniora-och-samhallsvetenskap.html.

Vetenskapsrådet. "DIGARV—Om DIGARV." Accessed March 2, 2021. https://www.digarv.se/om-digarv/.

Vetenskapsrådet. "Forskning & innovation för framtiden: Gemensam analys som underlag till regeringens forskningspolitik från Energimyndigheten, Formas, Forte, Rymdstyrelsen, Vetenskapsrådet och Vinnova." 2019. https://www.vinnova.se/globalassets/2019-02855-gemensamt-inspel.pdf.

Vetenskapsrådet. "Forskningsinfrastrukturdagen 2018." 2018. https://www.vr.se/aktuellt/evenemang/alla-evenemang/kalendariehandelser/2018-06-26-forskningsinfrastrukturdagen-2018.html.

Vetenskapsrådet. "Forskningsinfrastrukturdagen 2018 (Video)." October 16, 2018. https://www.youtube.com/watch?v=XXURjBbaPzI.

Vetenskapsrådet. "Natural and Engineering Sciences." October 31, 2017. https://www.vr.se/english/applying-for-funding/decisions/2017-10-31-natural-and-engineering-sciences.html.

Vetenskapsrådet. "Research Infrastructure." September 20, 2017. https://www.vr.se/english/applying-for-funding/decisions/2017-09-20-research-infrastructure.html.

Vetenskapsrådet. "Research Overview 2019: Humanities and Social Sciences." 2019. https://www.vr.se/english/analysis/reports/our-reports/2019-05-10-research-overview-2019-humanities-and-social-sciences.html.

Vetenskapsrådet. "The Swedish Research Council's Guide to Infrastructures 2012." 2012. https://www.vr.se/download/18.2412c5311624176023d2551b/1529480559612/Guide-Infrastructures_VR_2012.pdf.

Vetenskapsrådet. "Vetenskapsrådets årsredovisning 2016." 2017. https://www.vr.se/download/18.11999f69161e921384fae7cd/1555332001614/Vetenskapsradet_arsredovisning_2016.pdf.

Vetenskapsrådet. "Vetenskapsrådets guide till infrastrukturen 2014." 2015. https://www.vr.se/download/18.2412c5311624176023d25bff/1556010583275/VRs-guide-till-infrastrukturen-2014_VR_2015.pdf.

Vickery, Margaret Birney. *Landscape and Infrastructure: Reimagining the Pastoral Paradigm for the Twenty-First Century*. London: Bloomsbury Academic, 2020.

Viseu, Ana. "Integration of Social Science into Research is Crucial." *Nature* 525, no. 291 (2015): 291.

Vleuten, Erik van der. "Large Technical Systems." In *A Companion to the Philosophy of Technology*, edited by Jan Kyrre Berg Olsen, Stig Andur Pedersen, and Vincent F. Hendricks, 218–222. Chichester, UK: Wiley-Blackwell, 2009.

Wakkary, Ron. "Nomadic Practices: A Posthuman Theory for Knowing Design." *International Journal of Design* 14, no. 3 (2020): 117–128.

Walsh, Bryan. "Your Data Is Dirty: The Carbon Price of Cloud Computing." *Time*, April 2, 2014.

Walsh, J. "Behaviorial Sciences: The View at the Center for Advanced Study." *Science* 169, no. 3946 (1970): 654–658.

Warde, Paul, Libby Robin, and Sverker Sörlin. *The Environment: A History of the Idea*. Baltimore: Johns Hopkins University Press, 2018.

WASP-HS. "This Is: The Wallenberg AI, Autonomous Systems and Software Program—Humanities and Society. 2023." 2023. https://wasp-hs.org/wp-content/uploads/2024/01/This-is-WASP-HS-2023-Webb.pdf.

Wasserman, Sarah. "Ralph Ellison, Chester Himes, and the Persistence of Urban Forms." *PMLA* 135, no. 3 (2020): 530–545.

Waterston, Alisse. "Imagining World Solidarities for a Livable Future." *Kritisk etnografi: Swedish Journal of Anthropology* 3, no. 1 (2020): 33–42.

Watkinson, Charles, and Melissa Pitts. "Re-Envisioning Humanities Infrastructure." *Inside Higher Ed*, February 21, 2021.

Weigel, Moira. "Silicon Valley's Sixty-Year Love Affair with the Word 'Tool.'" *The New Yorker*, April 11, 2018.

Weizman, Eyal. *Forensic Architecture: Violence at the Threshold of Detectability*. New York: Zone Books, 2017.

Wenzel, Jennifer. "Turning over a New Leaf: Fanonian Humanism and Environmental Justice." In *The Routledge Companion to the Environmental Humanities*, edited by Ursula K. Heise, Jon Christensen, and Michelle Niemann, 165–173. New York: Routledge, 2017.

Wershler, Darren, Lori Emerson, and Jussi Parikka. *The Lab Book: Situated Practices in Media Studies*. Minneapolis: University of Minnesota Press, 2021.

Wesleyan University. "About the Center, Center for the Humanities—Wesleyan University." Accessed August 21, 2018. http://www.wesleyan.edu/humanities/about.html.

Whitehead, Anne, Angela Woods, Sarah Atkinson, Jane Macnaughton, and Jennifer Richards, eds. *The Edinburgh Companion to the Critical Medical Humanities*. Edinburgh: Edinburgh University Press, 2016.

Wikipedia. "List of Public Works Ministries." Wikipedia, April 11, 2023. https://en.wikipedia.org/w/index.php?title=List_of_public_works_ministries&oldid=1149260768.

Williams, Jeffrey J. "The New Humanities." *The Chronicle of Higher Education*, November 14, 2019.

Williams, Raymond. *Keywords: A Vocabulary of Culture and Society*. Rev. ed. New York: Oxford University Press, 1983.

Williams, Rosalind. "Infrastructure as Lived Experience (Filmed Keynote Lecture at Landscape Infrastructure Symposium)." Harvard GSD, March 23, 2012. https://www.youtube.com/watch?v=MfjkQjwoVHs.

Winn, James. "Interpretation as Performance: Promoting Dialogue between Scholars and Artists." In *The Institutional Impact of Institutes, University of California Humanities Research Institute, April 29.30, 1988—Papers for Discussion*, edited by Ralph Cohen, 1–5. Irvine: University of California Humanities Research Institute, 1988.

Winter, Deena. "Quietly, Gradually, George Floyd Square Is Open to Traffic for the First Time since His Murder." *Minnesota Reformer*, July 22, 2021. https://minnesotareformer.com/2021/07/22/quietly-gradually-george-floyd-square-is-open-to-traffic-for-the-first-time-since-his-murder/.

Wisselgren, Per. "Vetenskap och /eller politik?: Om gränsteorier och utredningsväsendets vetenskapshistoria." In *Mångsysslare och gränsöverskridare: 13 uppsatser i idéhistoria*, edited by Bosse Sundin and Maria Göransdotter, 103–119. Umeå: Umeå universitet, 2008.

Wittrock, Björn. "A Brief History of Institutes for Advanced Study." In "CAS Oslo 1992–2002: Advanced Study in a Norwegian Context," Oslo: Centre for Advanced Study, 2002.

Wolfe, Cary. *What Is Posthumanism?* Minneapolis: University of Minnesota Press, 2010.

Wolk, Sanna. "Sveriges universitetslärare: Politikerna försöker kidnappa svenska universitet och högskolor." *Altinget*, May 16, 2023. https://www.altinget.se/artikel/sveriges-universitetslarare-politikerna-forsoker-kidnappa-svenska-universitet-och-hogskolor.

Womersley, John. "More Effective Coordination of National Roadmaps." Presented at the InRoad Engagement Workshop, 2018. http://web.archive.org/web/20200907063937/https://www.inroad.eu/wp-content/uploads/2018/01/01_Keynote_John_Womersley.pdf.

Woodward, Kathleen. "Power and Place." In *The Institutional Impact of Institutes, University of California Humanities Research Institute, April 29.30, 1988—Papers for*

Discussion, edited by Ralph Cohen, 1–4. Irvine: University of California Humanities Research Institute, 1988.

Woodward, Kathleen. "The Future of the Humanities—in the Present and in Public." *Daedalus* 138, no. 1 (2009): 110–123.

World Trade Center Stockholm. "Conference Facilities | WTC." Accessed April 2, 2023. https://www.wtc.se/en/conference--restaurant/conference/.

Wormbs, Nina. "Därför måste vi våga moralisera i klimatfrågan." *Dagens Nyheter*, August 23, 2018.

Wright, Hannah. "Understanding Green Infrastructure: The Development of a Contested Concept in England." *Local Environment* 16, no. 10 (November 1, 2011): 1003–1019.

Yale University (Office of the Provost). "Exploring the Possibilities of Artificial Intelligence at Yale." Office of the Provost, January 17, 2024. https://provost.yale.edu/news/exploring-possibilities-artificial-intelligence-yale.

Yates, JoAnne, and Craig N. Murphy. *Engineering Rules: Global Standard Setting since 1880*. Baltimore: John Hopkins University Press, 2019.

Yergeau, M. Remi. "Disability Hacktivism." In *Hacking the Classroom: Eight Perspectives*, curated by Jentery Sayers and Mary Hocks. *Computers and Composition Online*, Spring 2014. http://cconlinejournal.org/hacking/#yergeau.

Yu, Betty. "Robotaxis Halt Traffic in San Francisco's North Beach Day after Expansion Approval (CBS San Francisco)." CBS San Francisco, August 13, 2023.

Zabusky, Stacia E. *Launching Europe: An Ethnography of European Cooperation in Space Science*. Princeton, NJ: Princeton University Press, 1995.

Zebrowski, Chris, and Daniel Sage. "Resilience and Critical Infrastructure: Origins, Theories, and Critiques." In *The Palgrave Handbook of Security, Risk and Intelligence*, edited by Robert Dover, Huw Dylan, and Michael S. Goodman, 117–135. London: Palgrave Macmillan UK, 2017.

Zeller, Thomas. "Aiming for Control, Haunted by Its Failure: Towards an Envirotechnical Understanding of Infrastructures." *Global Environment* 10, no. 1 (2017): 202–228.

Index

Bold page numbers refer to figures